I0474090

"Oil for the Lamps of China"–Beijing's 21st-Century Search for Energy

Bernard D. Cole

McNair Paper 67

INSTITUTE FOR NATIONAL STRATEGIC STUDIES

NATIONAL DEFENSE UNIVERSITY

WASHINGTON, D.C.

2003

The opinions, conclusions, and recommendations expressed or implied within are those of the contributors and do not necessarily reflect the views of the Department of Defense or any other agency of the Federal Government. This publication is cleared for public release; distribution unlimited.

Portions of this work may be quoted or reprinted without further permission, with credit to the Institute for National Strategic Studies. A courtesy copy of any reviews and tearsheets would be appreciated.

Contents

Figures

Introduction

In 1933, Alice Tisdale Hobart, wife of the Standard Oil Company of New Jersey manager in Nanking, published *Oil for the Lamps of China*.[1] Hobart had traveled widely in China and proved to be a very observant imperialist. Her fictional account of her experiences, not surprisingly, focused on the role played by Western businessmen, especially those engaged in importing and selling petroleum products. One thread that runs through her work is Chinese dependence on foreign sources of energy supplies, which remains the case today. This dependence on foreign-controlled sources means that Beijing's efforts to ensure the availability of energy resources adequate to fuel the nation's economic growth have important national security implications.

China is the second largest energy-consuming country in the world after the United States.[2] Beijing is determined to maintain continued economic growth after a quarter-century of remarkably high economic performance in the face of a population of 1.3 billion and a changing political system. The Organization for Economic Cooperation and Development (OECD) estimates that the average annual economic growth rate for China, including Hong Kong, will be 5.6 percent for the period 1995 to 2020, compared to 8.5 percent for the period 1971 to 1995.[3] China could be the largest economy in the world by 2050, in terms of purchasing power parity, with a gross domestic product slightly less than half of that of the 30 OECD member-nations combined. In a best-case prognosis, China has the strong potential to rank as at least the second largest world economy by the new century's midpoint. Economic expansion of that magnitude means China's domestic energy demand over the next 20 years would grow at a rate that is more than that of any other nation.[4]

China's economic growth has been accompanied by energy consumption growth averaging 5 percent per year—with electricity growth of 8 percent annually.[5] China was responsible for 9.6 percent of global energy consumption in 1997, a figure projected to grow to 16.1 percent by 2020. Coal accounted for approximately 74 percent of the country's energy production in 2000. In the same year, petroleum accounted for 18 percent, hydropower 5 percent, natural gas 2 percent, and nuclear power less than 1 percent. Total energy consumption in 2000 included 71.3 percent produced from coal, 21 percent from oil, 5 percent from hydropower, and 2 percent from natural gas.[6]

China became dependent on international sources of petroleum in 1993 and on all forms of international energy supplies in 1996, a dependence that continues to increase. Reliance on non-domestic energy sources is exacerbated by the fact that 10 percent of domestic production comes from offshore wells that are almost all the product of joint ventures with foreign companies.

In May 2000, Beijing announced a long-range comprehensive plan for its energy sector that includes development of thermal, nuclear, and hydroelectric power, in addition to oil, coal, natural gas, "and other new and renewable energy sources," including geothermal, solar, biomass, wind, and tidal energy. This plan also lines up with the current drive to develop China's western regions economically, with coal, hydropower, and the oil and gas industries all focusing in the west.[7]

The reliance on foreign energy sources also complicates this plan's chance for success; the deputy director of the State Economic and Trade Commission (SETC) has estimated that "China will have to rely on international markets for 50 percent of its oil supply in 2020."[8] In December 2000, senior industry spokesmen stated that "Beijing had no plans to depart from a schedule freeing up the country's oil products market to foreign investors."[9]

Beijing recognizes the strategic implications of national reliance on foreign energy sources and is attempting to lessen this reliance. China is the world's fifth largest petroleum producer but imported more than 18 percent of its petroleum consumption in 1999.[10] It imported twice as much oil in 2000 as it did in 1999, and 15 percent more in 2002 than in 2001.[11] The demand for imported petroleum will continue increasing, assuming continued economic growth, with imported oil contributing as much as 40 percent of all petroleum requirements by 2010 and with as much as 8 million barrels a day imported by China in 2020.[12] Maintaining

Figure 1. Energy Consumption per Person: China and the United States

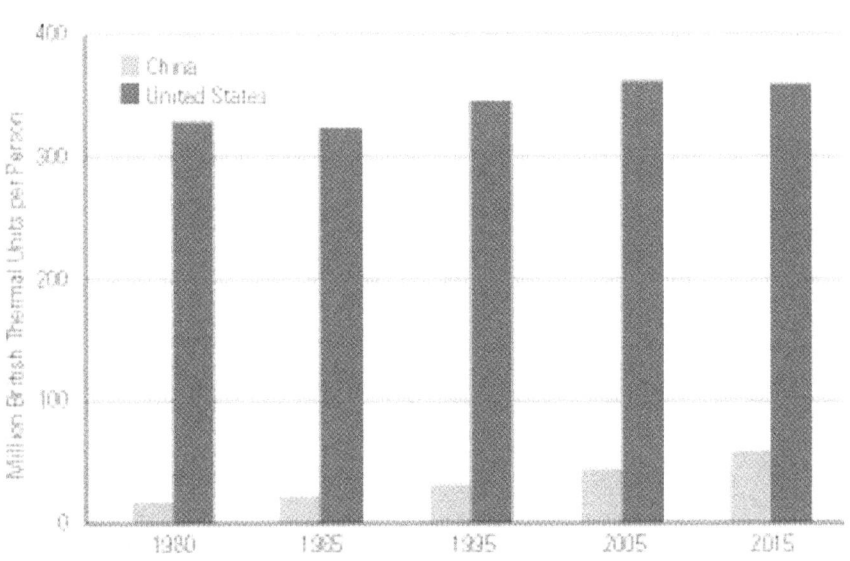

Source: U.S. Department of Energy, Energy Information Administration, Main Products of the Office of Energy Markets and End Use, Country Analysis Briefs, "China: An Energy Sector Overview," accessed at <http://www.eia. doe.gov/emeu/cabs/china/enerpcap.gif>

the rate of economic growth to which the Chinese people have become accustomed is key to sustaining the legitimacy of the current political system in China as its ideological basis erodes.

A further complicating factor is Beijing's campaign to privatize many important sectors of the economy hitherto dominated by the state. The power industry is one such sector; the government has announced its determination to end monopoly, enhance competition, and introduce competitive pricing in this industry.[13] National authorities undoubtedly are determined to conduct this campaign and recognize the inevitable social and political fallout of privatization. They are equally determined to change China's economic character while retaining the Chinese Communist Party (CCP) in power, but numerous statements by Beijing officials and Chinese academics indicate government flexibility in the face of economic priorities.

This essay addresses Chinese energy dependence and future policy options. An introductory review of Beijing's current energy situation

Figure 2. China's Energy Production and Consumption, 1980–2015

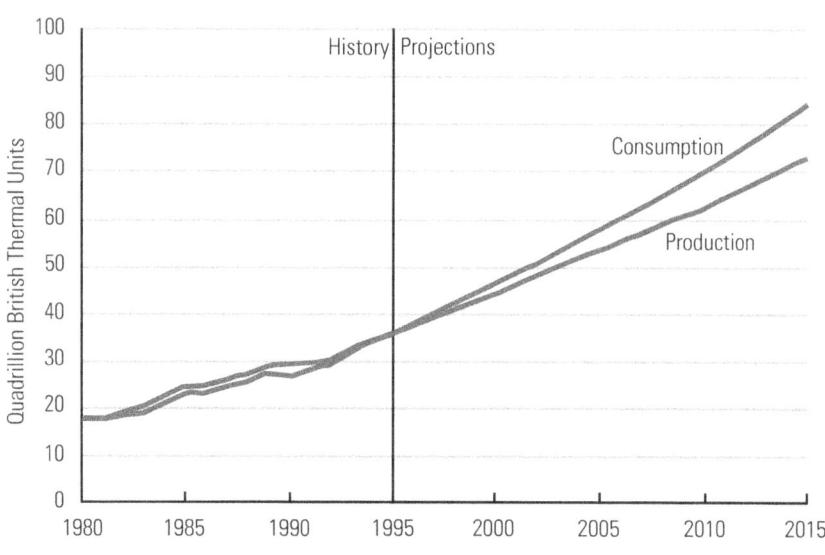

Source: U.S. Department of Energy, Energy Information Administration, Main Products of the Office of Energy Markets and End Use, Country Analysis Briefs, "China: An Energy Sector Overview," accessed at <http://www.eia. doe.gov/emeu/cabs/china/energy.gif>

is followed by a discussion of the government organizations that still dominate the energy sector, despite China's privatization campaign. Next are reviews of China's dominant coal industry, current oil production and imports, and natural gas sector. Other power sources are discussed, followed by a review of the electrical power distribution system envisioned by Beijing.

Environmental concerns associated with energy production are addressed, as is the effectiveness of Beijing's Ninth and Tenth 5-Year Plans and future energy policies. Defense of energy resources is reviewed, with a focus on both military and civilian police resources and apparent plans. The conclusion focuses on China's potential to solve its energy questions in the new century.[14]

Chapter One

Energy Sector Organization

T he energy industry in China to a significant degree exemplifies
the post-1978 development of "socialism with Chinese characteristics," including extensive privatization and ascendancy of the
profit motive.[15]

The *socialist market economic system* is defined as "socialist public
ownership as the mainstay [of the economic system with] diversified systems of economic ownership developing" simultaneously. The process is
still ongoing, with remaining problems in the growing population, poor
workforce quality, deteriorating ecology, and shortage of natural resources.[16] The implications of applying essentially free market ownership
to the energy sector will almost certainly make development of that sector
more dynamic and less susceptible to central planning.

An extensive restructuring of the energy industry in 1993–1994,
when the domestic oil market was allowed to adopt a more regulatory
than tightly, centrally controlled posture, was an early but unsuccessful
attempt to resolve these problems. Beijing is still moving down a quarter-
century path of privatization and deregulation of the energy sector, albeit
at a measured pace.[17]

A more extensive reorganization was implemented in 1998,
when the government restructured state-owned assets into two vertically
integrated firms; this plan recognizes market mechanisms while retaining at the "commanding heights" of the energy sector at least nominally
state-owned enterprises. The reorganization provided for a transfer of
assets between the two firms, transforming them into regional entities.
The China National Petroleum Corporation (CNPC) was assigned re-
sponsibilities mostly in China's north and west and the China Petroleum
and Chemical Corporation (Sinopec) in the south. Sinopec transferred
four northern refineries to CNPC, which transferred eight southern oilfields to Sinopec. CNPC still has more than two-thirds of China's crude

oil production capacity, while Sinopec controls more than half of China's refining capacity.

Furthermore, Sinopec is the primary importing company for crude oil, importing approximately 80 percent of the national total in 2001.[18] The firms have spun off or eliminated several ancillary activities that were unprofitable; many of these, such as housing units and hospitals, have obvious social implications, since their loss to workers and former workers no doubt exacerbated the labor unrest that has been a feature of Beijing's campaign to privatize and restructure state-owned enterprises (SOEs).

Other major state-sector firms in China include the China National Offshore Oil Corporation (CNOOC), which handles offshore exploration and production and accounts for roughly 10 percent of China's domestic crude production, and China National Star Petroleum (CNSP), created in 1997. CNOOC had planned an initial public offering (IPO) on the New York Stock Exchange in late 1999, but it was cancelled after the company failed to agree with its underwriters on an opening share price.

The three largest concerns, Sinopec, CNPC, and CNOOC, have all made IPOs of stock, attracting billions of dollars of foreign capital. Two unusual features marked these stock offerings. First, they all involved significant, albeit minority, holdings in the companies by foreign concerns. Second, although foreign holdings in important Chinese companies are not uncommon, such holdings in the vital energy sector may allow foreign investors a significant voice in company decisionmaking.[19]

The 1998 reorganization aimed to increase efficiency and profit, but also to strengthen state control over the domestic oil sector. In late 1999, CNPC set up a holding company, PetroChina, for the purpose of raising money on international capital markets. PetroChina includes most CNPC productive assets but excludes its network of employee-support functions and some controversial projects such as its holdings in Sudan.[20] An IPO of a minority stake in PetroChina on the New York and Hong Kong stock exchanges took place April 7, 2000, though the size of the offering had to be scaled back due to a lack of interest on the part of institutional investors.

Three types of organizations currently govern China's energy sector (figure 3). Ministry-level corporations run the highly centralized petroleum and nuclear industries (reporting to the State Council, headed by the nation's premier, with a vice premier for energy issues), while energy subministries and affiliated national corporations run the less centralized electric power and coal industries.

- The State Planning Commission has ultimate authority for energy project approval, budget allocations, and financing arrangements.
- The State Science and Technology Commission and the State Economic and Trade Commission are also involved with energy industry development.
- The China National Energy Investment Corporation oversees major investment loans for the energy sector.
- The China National Petroleum Corporation is responsible for all on-shore upstream oil and gas operations, including shallow-water areas. In the past few years, CNPC has begun transformation into a multi-national integrated oil company, establishing subsidiaries and acquiring overseas acreage and refineries in pursuit of export markets.
- The China National Offshore Oil Corporation was established in 1982 to explore China's offshore petroleum resources. The corporation has four regional subsidiaries (Bohai, East China Sea, Nanhai East, and Nanhai West) and several specialized subsidiaries.
- China established a third state oil company, the China National Star Petroleum, in January 1997. The company has been authorized by the central government to launch several exploration ventures with foreign companies.
- The China Petroleum and Chemical Corporation (Sinopec) is responsible for petroleum processing and product distribution; it controls production facilities for 90 percent of China's refined oil products and over 75 percent of its petrochemicals.
- The China National Chemicals Import and Export Corporation (Sinochem) is primarily involved in imports and exports of crude oil, petroleum products, and natural gas.
- The Ministry of Coal Industry allocates national coal production and coordinates production activities by central-government-controlled mines (about 45 percent of nationwide production) and local mines (collective or privately owned mines as well as state-owned mines operated at the provincial, prefectural, or county level).
- The Ministry of Electric Power regulates power production.
- The State Power Corporation was established in 1997 to handle business aspects of the industry, with generation and transmission the responsibility of regional subsidiaries.

Figure 3. Organization of China's Energy Sector

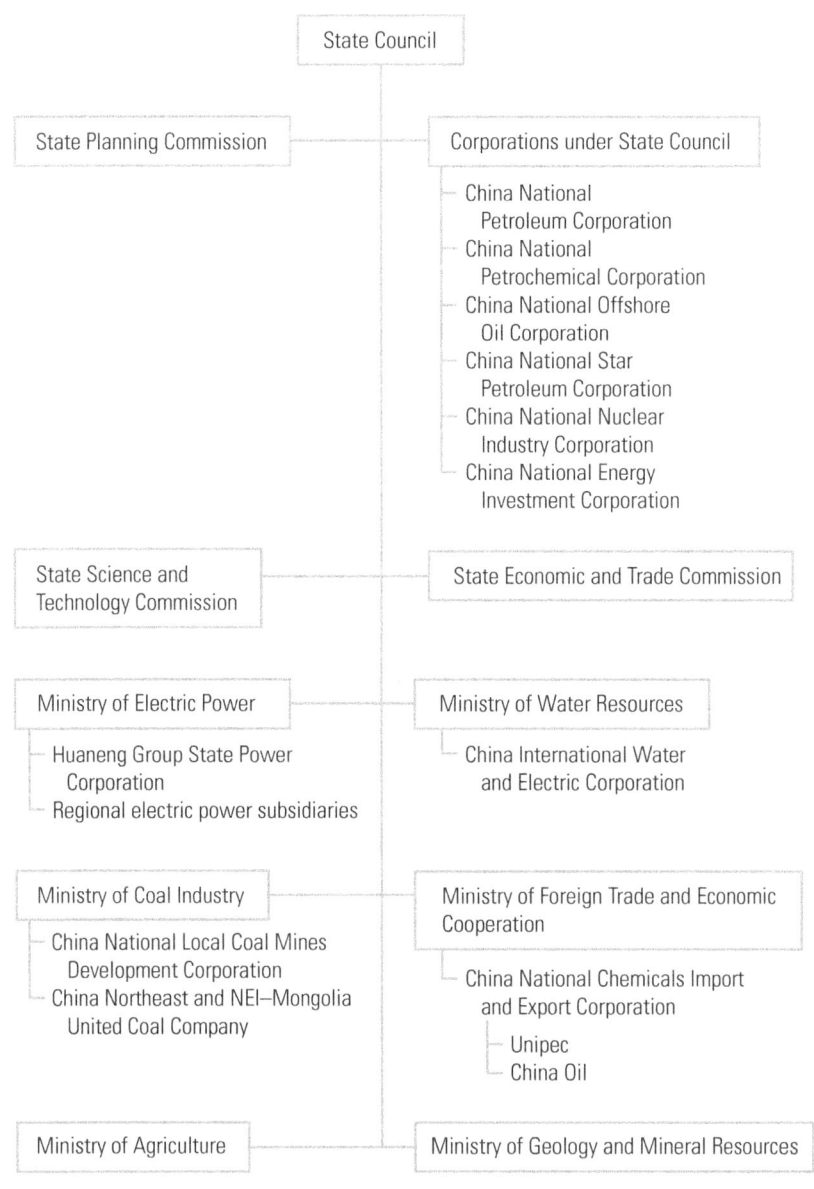

Source: U.S. Department of Energy, Energy Information Administration, Main Products of the Office of Energy Markets and End Use, Country Analysis Briefs, "China: An Energy Sector Overview," accessed at <http://www.eia/ doe.gov/emeu/cabs/china/org.gif>

- The Ministry of Water Resources is concerned with China's hydropower production.
- The China National Nuclear Corporation is a conglomerate of more than 200 enterprises and institutions concerned with China's nuclear power production and waste disposal facilities.

This infrastructure recently has been augmented by a national plan to design a strategy of sustainable energy development, which will include establishing forums for experts, training government officials, and conducting surveys and investigations aimed toward further energy supply discoveries.[21] The effort is being supervised by the State Development Planning Commission and the China International Center for Economic and Technological Exchange. It also has gained the endorsement of the United Nations Development Program (UNDP), which has allocated $300,000 for the project. Tibet is included in this plan, with developmental emphasis in that province placed on hydropower and solar power resources.

This strategy is a subset of a Beijing plan for nationwide development of the industrial sector and embodies the privatization believed necessary for China's economic ambitions. An important goal of the plan is to "improve the investment environment for the power industry and upgrade power facilities" to satisfy demand estimated to grow at an annual rate of at least 4 percent through 2005. The plan aims to include natural gas, hydropower, and nuclear power projects, in addition to increasing the efficiency of the national power-generating grid, with particular attention to linking power availability in the western and eastern parts of the country. Four targets of development are described: installed capacity goals, power grids linked and "more rational," increased environmental protection, and power supply available to all Chinese villages.[22]

Much of this effort is aimed at alternative—that is, non-hydrocarbon—energy source development. Current Chinese planning also emphasizes foreign investment, especially in the riskier plans, such as those involving solar, wind-electric, marine energy, terrestrial heat, and biological energy production. The Asian Development Bank has extended loans for developing biological and solar energy as part of "59 international cooperation projects involving new and renewable energy development." The World Bank has approved its largest-ever renewable energy loan and its first renewable energy loan for China, $100 million plus a $35 million grant from the Global Environment Facility. Funds are to be used to develop wind power in Inner Mongolia, Hebei, Fujian, and Shanghai and to develop solar power in isolated rural areas in the northwest of the country.[23]

Chapter Two

The Coal Industry

Despite efforts to diversify energy sources, China is expected to reduce its reliance on coal by only 10 percent by 2020: coal is readily available, under Beijing's direct control, and influences a large portion of the country's labor force. A 10 percent reduction will result in a major change in that force but will likely be balanced economically by advantages gained by shifting to cleaner, more efficient energy sources.[24] Although other energy sources are expected to grow, China's main energy source for the foreseeable future will remain its very extensive coal reserves (see figure 4). Coal's 75 percent share of China's primary energy supplies produced over 62 percent of commercial energy consumption. According to the Coal Industry Ministry, coal remains "the indispensable energy source for China's economic development."[25]

China's estimated total coal resources are second only to those of the former Soviet Union, although it ranks third in the world in proven reserves (behind the United States and the former Soviet Union), due mainly to a lack of exploration. China has concentrated its resource development efforts on the higher-grade bituminous and anthracite coal that makes up more than half of its estimated 126 billion tons[26] of reserves; bituminous coal accounts for about 75 percent of annual production, and anthracite most of the rest.

Three-quarters of the electricity generated in China is coal-fired, and its coal industry is the world's largest, producing nearly 1.4 billion tons in 1996. The growth of coal production was originally forecast to reach about 1.5 billion tons by the year 2000 and 2.1 billion tons by 2010, but recent government moves to match production levels to slowing consumer demand resulted in approximately 1.1 billion tons produced in 2000.

As of late 1999, about 39 percent of Chinese coal consumption was consumed in power stations, 14 percent for coking, 10 percent for

Figure 4. Energy Production by Source, 1980–2015

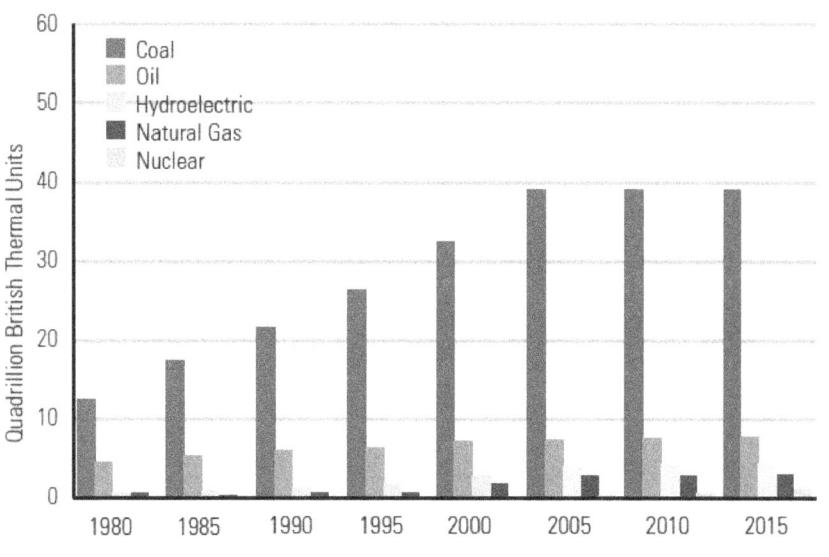

Source: U.S. Department of Energy, Energy Information Administration, Main Products of the Office of Energy Markets and End Use, Country Analysis Briefs, "China: An Energy Sector Overview," accessed at <http://www.eia.doe.gov/emeu/cabs/china/prodfuel.gif>

domestic and residential use, 1 percent for rail, and the rest by other industries such as chemical, cement, ceramics, and glass-making plants. By contrast, the United States burns 87 percent of its coal in power utilities.[27] Plans for power station development provide for coal as the primary fuel.

In 2000, China consumed an estimated 1.68 billion short tons of coal, by far the largest national consumption in the world, with the United States in second place at 1.09 billion short tons. This figure is expected to grow to 3.54 tons by 2020, dramatically increasing China's position as the world's largest consumer of coal.[28] Coal exports also continue to increase, to almost 21 million tons during the first half of 2000, a 45-percent increase over the same period in 1999. Beijing still faces an oversupply of coal, a problem it is trying to solve by closing thousands of small-scale coal mining operations.[29]

As a result of the extensive reorganization that began in the late 1990s, less than half of China's coal production is now centrally administered, with the majority produced by either local state-owned mines or rural collectives. In the early 1980s, the central government began to encourage the development of small coal mines in rural areas; many of these mines are individually owned. Amendments to China's Mineral Resources Law, effective January 1997, provide a comprehensive legal framework for exploration and exploitation and could encourage foreign investment.

Northern China, especially Shanxi Province, contains most of China's easily accessible coal and virtually all of the large state-owned mines. Coal from southern mines tends to be higher in sulfur and ash, therefore unsuitable for many applications. Regional imbalances between coal supply and demand require the transporting of large quantities of coal, generally from northern areas to the south and east.

In fact, coal accounts for a larger percentage of freight than any other commodity in China. More than 50 percent of the coal is shipped by rail, and 60 percent of rail transport is tied up in transporting coal. The major north-south rail lines are Zhengzhou-Wuhan and Xuzhou-Nanjing, and the primary east-west line is Datong-Qinhuangdoa. Transportation bottlenecks are a major problem in the coal industry, with the high content of dirt and rocks included in shipments (only 20 to 25 percent of coal is now washed) further taxing the system. Recommended solutions include rail expansion projects and alternatives such as coal pipelines, liquefaction, and coal-by-wire (locating power plants at mine mouths).

The China National Coal Import and Export Corporation is the primary Chinese partner for foreign investors in the coal sector; this corporation is overseeing a process more open to foreign investment, particularly in modernizing existing large-scale mines and the developing new ones. Areas of interest in foreign investment concentrate on new technologies only recently introduced in China or with environmental benefit, including the coal liquefaction, coal bed methane production, and slurry pipeline transportation projects. These projects also promise to increase the efficiency of the transportation system.[30] In addition to collocating coal-fired power plants with large mines, other technological improvements are being undertaken, including the first small-scale projects for coal gasification, and a coal slurry pipeline to transport coal to the port of Qingdao.

China is an increasingly important participant in the international coal exporting market. Beijing tripled its coal exports between 1980 and 1990, as newly built coal washing, rail, and port facilities made more of its high-quality coal available for export. Beijing announced net coal exports for 1998 of 35 million tons, with sales primarily to South and North Korea and to Japan. Japan not only is China's biggest customer but also has become a partner in the industry, providing loans for the improvement of railroads and ports for the overseas transport of coal from Shanxi Province. A 5-year agreement for Chinese energy exports to Japan was negotiated in December 2000.[31]

As China commits itself to further economic reform and increased involvement in the global economy, its coal industry faces major challenges of rationalization and structural reform. The government is implementing major reforms to reduce the effects of the current oversupply problem, at least in the short term. Large state-owned coal mines have experienced buildups of unused inventories, and many are operating at a financial loss. A large number of small, unlicensed mines also contribute to this situation.

Coal usage is also very susceptible to swings in the Chinese economy. This has influenced the recent overhaul of the state coal administration, radical cuts in production, the restructuring of key state-owned coal mines, and the planned closure of over 25,000 small mines. In 1998, the government launched "a campaign to shut thousands of coal mines, especially small and dangerous ones, to avoid overproduction, falling prices, and to improve the industry's safety record."[32]

Poor Safety Record

One report, in early April 2002, stated that 59,000 of an estimated 82,000 illegal small coal mines have been closed, but another claimed that 430,000 had ceased operation. Neither figure is likely definitive, given the central government's apparent lack of an effective regulatory infrastructure in this area.[33] This campaign appears to have achieved mixed results; as a result of the closures, depressed local coal prices have started to recover, but supply still outpaces demand.

The government reported that 5,395 miners died in mining-related accidents in 2001, but other estimates range to up to 10,000 miners dying annually with perhaps thousands more dying each year from lung diseases.[34] Most of these occurred in the many small, unregulated mines.

Ninety-two miners died in a Jiangsu Province gas explosion in July 2001, while in September, at least nine miners died in an explosion at an illegal coal mine; in November, 37 died in two coal mine blasts in a mine in Podi, Shanxi Province, that had been ordered closed and was "operating without government permission." That same month, another explosion claimed at least 14 miners.[35] The operators of the Podi mine were quickly tried and sentenced to prison terms, but the accidents continue; in

December 2001 alone, 11 died in Hunan, 15 in Henan, 8 in Liaoning, and over 20 in Jiangxi Provinces.[36]

The national government had turned its full attention to attempting to regulate the industry even before this latest series of disasters. In addition to directing the closure of many small mines, Beijing issued a completely revised set of coal mine safety regulations, which took effect December 1, 2000.[37] These include "supervision of safety in coal mines" by the Coal Mining Safety Supervisory Organization as an arm of the State Council. Mirror safety organizations also exist at provincial, autonomous regional, and autonomous municipal levels (Article 8). Governmental frustration with coal entrepreneurs and mine owners is indicated in Article 6, which states that "coal mine safety supervision shall rely on miners and union organizations." The regulations include detailed requirements for the construction, equipping, and operation of coal mines—and provide penalties, including mine closure, for noncompliance.[38] The leadership also began paying personal attention to the problem, as then Vice Premier Wu Bangguo visited coal miners in Shanxi Province, and Vice Premier Wen Jaibao did the same in Liaoning.[39]

China's mining safety record remains horrendous, however, with a continuing series of disasters that the government seems unable to halt. Notwithstanding the new regulations and "billions of yuan put into improving mine safety," the government reported a recorded 556 coal mine accidents in just the first quarter of 2002, causing 994 deaths.[40] Reported accidents in early 2003 indicated that a solution to the mining accident problem remained lacking. Eight miners died in Jilin Province, 38 in Heilongjiang, 8 in Henan, and at least 38 in Guizhou.[41]

Prospects for Coal

Despite these problems of inefficiency and safety, China's coal output continues to rise, and exports are profitable.[42] Coal will almost certainly remain the dominant energy source in China for the foreseeable future, despite the strenuous efforts of the authorities to diversify the energy mix.

While the infrastructure for coal transport, including rail and ports, has improved significantly over the past 8 years, it remains inadequate to meet current demand.[43] In fact, imports of coal have risen in coastal regions, due to the inefficiencies of domestic rail transport. However, the completion of the Shuo-huang railway (the second dedicated coal line) and the Huanghua coal terminal in 2003 will provide up to 60 million tons per year of new coal transport capacity in northern China. The first section of this railway began operation in December 2000; it is planned as part of a system to link the coalfields of the Ningxia Hui and Inner Mongolia Autonomous Regions to coastal China.[44]

China is not making the large investment necessary to build the infrastructure required to take efficient advantage of its huge reserves of coal, which will continue as its primary energy source. This situation is exacerbated by the scarcity of investment capital and water shortages that restrict efforts to improve the quality of the coal through the greater use of coal washing plants. China's relatively unsophisticated heavy industrial plant further means general satisfaction with low-grade coal, cheaper but both less efficient and more polluting. Hence, the resulting lack of demand for better quality coal and the absence of a fair pricing regime to reflect the value of coal quality constrain the modernization of the nation's coal industry as a whole.

While the high level of coal usage—both in real terms and as a percentage of energy production—will continue and even grow, how coal is utilized will change. Residential and domestic coal consumers typically contribute disproportionately to the high pollution levels. As more in this sector are provided with electricity and gas, they will burn less coal, lowering pollution. No such decrease in pollution seems apparent in the future of the industrial sector, however, where many boilers and plants have low thermal efficiency and lack modern pollution control equipment. National and provincial authorities will have to both mandate more stringent pollution limits and offer financial incentives to meet those limits.

Coal bed methane may provide one alternative, with reserves estimated at 30 to 35 trillion cubic meters (tcm) and production expected to reach 0.4 tcm by 2010.[45] China recently launched a project on the "Research and Application of Underground Vaporization of Coal" to attempt to increase recovery of this combustible gas for commercial use.[46]

Coal bed methane production is being developed under the aegis of the China United Coal Bed Methane Company signing agreements with 19 foreign companies, including American investors Atlantic Richfield and Texaco, the latter signing 3 production contracts. This joint effort will explore coalfield blocks in Inner Mongolia, and Shaanxi, Shanxi, and Anhui Provinces.[47] Another joint venture project is planned for Yunnan Province.[48]

New technology may affect China's entire energy use profile in a dramatic fashion. Should the current U.S. program to develop direct carbon conversion bear fruit, the process would help to conserve precious fossil resources by allowing more power to be harnessed from the same amount of fuel. It would also improve the environment by significantly decreasing the amount of pollutants produced per kilowatt-hour of electrical energy generated, including decreasing carbon dioxide emissions.[49]

Chapter Three

The Oil Industry

P etroleum is the second largest energy source in China, with a domestic output of over 2,353 million barrels (bbl) of refined petroleum products in 2001.[50] About 1993, China's domestic oil production began to fall behind its growing oil consumption, and the country became a net importer of oil (see figure 5). Demand for petroleum is forecast to increase by 50 percent by 2020. Furthermore, dependence on foreign petroleum sources, more than 368 million bbl in 2000, will also increase, probably doubling to more than 735 million bbl by 2020, which will form approximately 50 percent of China's total petroleum consumption.[51]

Beijing is pursuing several avenues to reduce reliance on foreign petroleum supplies. These include maximizing sources on Chinese territory, both on- and offshore; securing foreign sources through acquisition both of production sources and facilities, as well as the product; creating a national strategic stockpile; and enhancing the physical security of stocks and sources on hand, all to be supervised by "a unified management body."[52]

China's oil and natural gas industries are almost exclusively government-owned, with the exception of a limited number of joint ventures. The main focus is on oil development, although the current 5-Year Plan seeks to take greater advantage of natural gas resources. Most oil is produced onshore by CNPC. The central government maintains active control over China's most productive fields, including the largest, at Daqing in the Songliao Basin and the Shengli and Liaohe fields in the Bohai Basin, all in northeastern China.

The China National Oil Development Corporation, a CNPC subsidiary, is the contracting agent for cooperation with foreign companies in the onshore oil industry. China's program to acquire interests in petroleum exploration and production abroad is led by CNPC, which holds oil

Figure 5. China's Oil Production and Consumption, 1980–2000

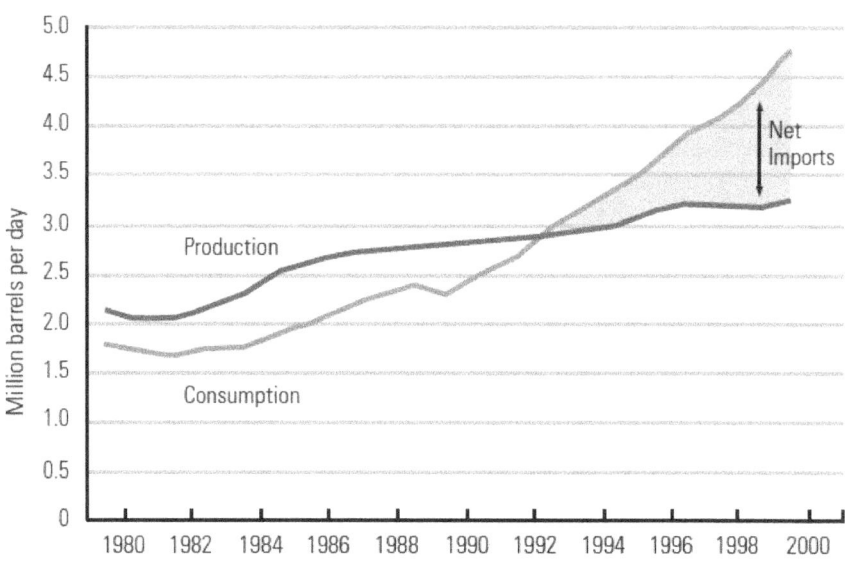

Source: U.S. Department of Energy, Energy Information Administration, Main Products of the Office of Energy Markets and End Use, Country Analysis Briefs, "China: An Energy Sector Overview," accessed at <http://www.eia.doe. gov/emeu/cabs/china/energy.gif>

concessions in Kazakhstan, Kyrgyzstan, Venezuela, Sudan, Iraq, and Peru. The Greater Nile Petroleum Operating Company, the Sudanese oil project in which CNPC owns a stake, began exports in August 1999. The CNPC concession in Iraq cannot be developed until the United Nations (UN) economic sanctions are lifted, at least to the extent of allowing foreign investment in Iraqi oil infrastructure.

China is also establishing energy relationships "all across the Middle East, Southeast Asia, Russia, Central Asia, Africa," and Latin America.[53] Beijing clearly is pointing the national search for energy to a massive global campaign, especially for petroleum. The ongoing reorganization of the energy sector is fostering a degree of privatization that undoubtedly makes expanded exploration attractive to the very large companies now dependent on profits for continued commercial viability.

Interestingly, Taiwan is trying to partner with Chinese companies searching for new oil sources. CNOOC signed an agreement with Taiwan's China Petroleum Corporation in January 2002 to search for oil in the Taiwan Strait.[54] The joint venture has been held up, however, by

Taiwan's Mainland Affairs Council; should that body approve the proposal, moreover, Beijing may well decide not to approve a Chinese company (and one that is primarily a government entity, at that) helping Taiwan overcome its lack of domestic petroleum by drilling in the waters between the island and the mainland. The economically logical business agreement may still fall victim to politics, although a May 2002 report stated that CNOOC and Taiwan's state-run China Petroleum Corporation had "signed a landmark agreement for joint exploration."[55]

CNOOC has traditionally sought foreign investment for offshore oil, which represents a relatively small share of China's oil industry. The company reportedly has interests in 45 maritime oil and gas properties in the Bohai, South China, and East China Seas. More recently, CNOOC has drawn interest from several foreign companies to its proffered deepwater blocks; an American company, Husky Oil, is the first to sign an agreement to explore one of these 12 blocks, in an area 100 kilometers (km) southeast of Hainan Island.[56]

As of late 1996, foreign commitments by Chinese energy companies totaled nearly $3 billion (almost 60 percent of total offshore exploration and development), including CNPC exploitation of oilfields in Russia, Pakistan, Kazakhstan, Indonesia, Myanmar, Egypt, Ecuador, Venezuela, Argentina, Iran, Iraq, Somalia, and Sudan, as well as power generation and refining projects in Bangladesh, Kuwait, Sudan, and Kenya.[57] In fact, CNPC president Ma Fucai has stated that the company hopes to emulate ExxonMobil, with "overseas production account[ing for] 60 to 70 percent" of profits.[58]

These projects, with a contract value of $750 million, are augmented by proposed pipelines with or in Russia, Turkmenistan, and Thailand, and possibly Pakistan, Afghanistan, and Burma.[59] Other Chinese projects (some with foreign partnership) have been established in Canada, Colombia, Malaysia, Mexico, Mongolia, Papua New Guinea, and the United States. China's first joint venture refinery, West Pacific Petrochemical Company (20 percent owned by Total SA of France), opened in late 1996.[60] Through 2000, China's investment plans emphasized upgrading and expanding existing refineries, in some cases to handle imported crude oil from the Middle East, rather than constructing new facilities. This policy probably reflects both Beijing's desire to increase the efficiency of its refinery infrastructure and the global surplus in refining capacity.

Sinopec produces most of China's refined petroleum products, while Sinochem dominates oil and gas trade, with import and export quantities determined by state planners. However, Sinochem also has partnerships with CNPC and the China International United Petroleum and Chemical Corporation. Some CNOOC production-sharing agreements also allow direct exports from offshore fields. Future import routes under discussion focus on land-based facilities, including oil and natural gas pipelines from Russia and Central Asia.

Official encouragement of cooperation with foreign companies is reflected in recent regulations increasing the number of Chinese companies authorized to negotiate, contract, and implement "joint onshore oil exploration projects."[61] This objective was further underlined in December 2001, when the nation's trade minister urged expanded use of foreign capital throughout the country's economy.[62]

The need for foreign capitalization and expertise reflects the weakness of China's indigenous capabilities in the energy sector. The sector's shortcomings are also indicated by Shanghai's search for foreign partners to build a "natural gas grid" in the city.[63] The city of Hangzhou, in Zhejiang Province, has also announced a plan to "turn itself into a clean 'natural gas city.'"[64] One of Shanghai's major projects is the conversion of city buses to gas fuel.[65]

China draws the majority of its overseas oil from the Middle East and Southwest Asia; its efforts to secure new foreign sources of petroleum are global in scope but have focused on Central Asia and Russia. Oilfield investment and pipeline construction are two important areas for Chinese companies and include projects in Azerbaijan, Pakistan, Uzbekistan, Kazakhstan, Kyrgyzstan, and Russia.[66]

The campaign to obtain energy resources from the former Soviet republics of Central Asia has occurred in the context of Beijing's diplomatic efforts to secure and expand Chinese influence in this area, historically of vital national interest to China. The primary vehicle of this effort has been the Shanghai Cooperative Organization organized in 1996 as the Shanghai Five by Beijing and Moscow; the other original member states were Kazakhstan, Kyrgyzstan, and Tajikistan. The organization's stated goals were to:

> strengthen mutual trust and good-neighborly relationship[s], increase effective cooperation of the member states in political, trade and economic, scientific and technological, cultural and educational fields, ensure peace, security and stability in the region, and push

forward the process of establishing a democratic, just, and rational international political and economic order.[67]

Five primary transnational pipelines to China have been discussed, although their actual construction will presumably depend on sufficient supplies and consumer demand:

- 1,865-km gas pipeline from Western Siberia to Shanshan, in Xinjiang Province; this $3.62 billion line would be able to transport 30 billion cubic meters (bcm) of natural gas to China annually.

- 3,370-km pipeline from Kazakhstan's Karachaganak gas and oilfield to Shanshan; this line would have a capacity of 25 bcm per year.

- 2,150-km pipeline from Turkmenistan to Shanshan, also with an annual capacity of 25 bcm, at a cost of approximately $4.7 billion.

- 2,416-km pipeline from Sakhalin to Shenyang, with an annual capacity of 5 bcm.

- a line from Irkutsk, in Siberia, to northern China (this plan is the most tentative of the five).[68]

In Kyrgyzstan, Beijing has approved a $300 million investment in a joint venture with the Kyrgyzneftegaz Company, to extract gas and oil from the country's south.[69] The main Chinese effort in Central Asia, however, has been in Kazakhstan, bordering on China's westernmost province, Xinjiang.[70] Kazakhstan, which may rank among the world's top six petroleum and natural gas producers by 2010, and China have signed a series of agreements to produce and export petroleum.

The most significant deal thus far has been the CNPC acquisition of a 60-percent stake in the Kazakh oil firm Aktobemunaigaz, based on a pledge to invest $4.6 billion in the company's development over the next 20 years.[71] Beijing has an ambitious long-range plan to build a pipeline from Kazakh natural gasfields to a central terminus in Xinjiang Province, whence a pipeline network will deliver the gas to central and eastern China, reaching to Shanghai. One Kazakh source has set a goal of exporting 50 million tons of oil a year to China.[72] To date, however, these contracts have led to very little pipeline construction; while Sino-Kazakh border disputes apparently have been resolved, weak economic indicators, political uncertainties, and severe corruption remain.

Beijing plans to meet this terminal with pipelines from Gansu, Szechuan, and Hebei provinces, thus distributing Central Asian energy supplies across much of China. Construction of the international portion of this pipeline was scheduled to begin in mid-2001, with the entire,

3,000-km-long network completed by 2008 at an estimated cost of $3.5 billion, but planning has halted, probably because of disagreements between Beijing and Almaty.[73] China's interest in Kazakh oil projects continues, however, with a focus on finding and exploiting new oil and gas deposits.[74]

CNPC friction with Kazakh authorities exists about the level of investment in the firm's projects, as well as other management issues. The senior CNPC representative for the Kazakh scheme, Zhang Chenwu, recently expressed reservations about the possibility of successfully completing the current pipeline project.[75] Political corruption on both sides of the Sino-Kazakh border is also a factor, as are governmental disputes within Kazakhstan, natural difficulties, and the slow development of Central Asian petroleum fields. These problems cast doubt on how soon China will be able to take advantage of this energy source in a major way.[76]

"Stabilize the East, Develop the West" is the current slogan in China's petroleum industry, which is applying enhanced oil recovery techniques to older fields and investing in promising areas of the West—in particular the remote Tarim Basin in the harsh environment of the Taklamakan Desert. A less remote but smaller target for development is the Turpan-Hami, or Tuha, Basin. A new 300-mile pipeline serving both of these areas was completed in 1997.

The South China Sea is surrounded by seven other nations, most of whom claim part or all of the sea's resources. The northern and western areas of the South China Sea are the scene of the most active offshore petroleum development; territorial disputes involving the Spratly Islands complicate offshore activities in the sea's central and southern areas. Since it is China's only contested area of known energy reserves, garnering those resources may rest directly on the ability of the People's Liberation Army Navy (PLAN) to enforce national claims.

Oil prospecting in the South China Sea dates back at least to Japanese efforts in the 1930s. Total 1998 production was over 1.3 million bbl per day, from proven reserves of approximately 7.5 billion bbl. Almost all of this came from the uncontested northern or southern areas of the sea.[77] There are no proven reserves for the Spratly or Paracel Islands in the sea's central area, and no commercially recoverable oil or gas has been discovered there, although efforts continue.[78] Geologists and analysts disagree on the presence (and recoverability) of petroleum reserves near these islands, with widely varying estimates.

The western sea area also promises to become a significant natural gas source, with 4 fields and 15 gas-bearing structures believed to exist.[79] These fields are expected to provide three bcm of gas by 2005, and six billion by 2010.[80]

Estimates of the petroleum reserves in the South China Sea range from Beijing's wildly optimistic 213 billion bbl (105 billion in the area of the Spratlys) to a U.S. estimate of 28 billion bbl (2.1 billion in the Spratlys).[81] A similar range of estimates exists for natural gas reserves, about which the Chinese are also optimistic, offering an estimate of more than 2,000 trillion cubic feet (tcf), while the U.S. estimate is a more modest 266 tcf. China's belief in these estimates is more important than their dubious accuracy, and Beijing's high expectations strengthen its determination to protect its sovereignty claims in the Spratly Islands.[82]

More than 368 million bbl of crude oil have been produced since the first well began pumping in 1990 and production expands annually as additional wells are installed, usually with participation by foreign companies. [83] The history of the Lufeng 22–1 oilfield, located in the northern South China Sea, illustrates the difficulties of recovering petroleum reserves from the area: the petroleum in these fields is typically located in high porosity sandstone and is waxy in texture, which requires extraordinary drilling and recovery methods. Additionally, the wells are located in a prime area of tropical cyclones that historically have had an average duration of 26 hours, generating a wave height of up to 8 meters. Force 6 winds and extremely rough seas are also common to the area. Specialized tankers have been designed and built to transport this field's product to the mainland—tankers that presumably would require protection from the PLAN in times of international crisis.[84]

Beijing's 1992 Law on the Territorial Sea and the Contiguous Zone brings it into direct contention with the other claimants to South China Sea resources: Vietnam, Malaysia, Indonesia, Brunei, and the Philippines; Taiwan makes the same claim as China. The 1992 law repeats Republic of China claims first made in the mid-1930s, including a U-shaped, dashed line encompassing almost all of the South China Sea, including ocean areas. This line implies that Beijing claims the entire sea as sovereign territory.

If upheld, this claim would have a major effect under the 1982 United National Convention on the Law of the Sea (UNCLOS), which most nations of the world—including China but not the United States—have signed. Not only would all the land formations that lie in the South

China Sea—primarily the Paracel and Spratly Islands (Xisha and Nansha in Chinese) and the Macclesfield Bank—be Chinese territory, but most of the very important natural gas and petroleum fields lying in the southernmost part of the sea would belong to China instead of to Brunei and Indonesia, which currently mine them.

The South China Sea also bears the main, crucial sea lines of communication (SLOCs) between East Asia and the oil-rich countries of Southwest Asia and the Middle East. Beijing's sovereignty over these SLOCs would be unacceptable to the United States, which insists on freedom of navigation through international waters. Many other nations would share American dissatisfaction with Chinese sovereignty over this sea, and a crisis situation would develop immediately, despite Beijing's protestations that its claimed ownership of these waters would not in any way affect the SLOCs.[85]

As a net crude oil importer, China's petroleum industry is focused on meeting domestic demand, but it still exports significant amounts of crude. The largest export customer by far is Japan, which imports 120,000 to 160,000 bbl per day of Daqing crude oil for generating electricity. This supply was called into question in early 1999, when CNPC informed the Japanese that it would no longer be available for export after deciding to refine it for sale on the domestic market rather than sell it at low world market prices. After Japanese protests to the State Bureau of Petroleum and Chemical Industries, CNPC agreed to continue supplies. The incident underlined the tension inherent in having state-owned firms operating with substantial independence—they still have to take the government's foreign policy concerns into account when making sales decisions.

An important development is China's decision to set extremely low quotas for crude imports—almost an import ban—starting in the second half of 1998, due to a supply glut and excess domestic refining capacity. Cheap imported oil was flooding the more prosperous coastal areas of China, causing stockpile increases and production cuts by domestic producers. The government responded by acting to protect domestic production. Crude oil import restrictions have largely been lifted, but restrictions on petroleum product imports are expected to continue in the near future.

Related to the import quota reduction is a continuing crackdown on petroleum smuggling, which has resulted in numerous criminal convictions. China's government also has begun a campaign to close down small-scale independent refineries, some of which had acted as conduits for smuggled oil. A Chinese government spokesman stated in February 2000 that all small refineries failing to meet the government's product quality standards would be shut down by the end of March 2000.

Approximately 90 percent of China's oil production capacity is located onshore. One complex alone, the Daqing fields in Heilongjiang Province, accounts for 1 million bbl per day of China's production, out of a total crude production of 3.2 million bbl per day. Daqing is a mature field, however, and is expected to show a declining output in future years. Government priorities focus on stabilizing production in the eastern regions of the country at current levels, increasing production in new fields in the west, and developing the infrastructure required to deliver western oil and gas to consumers in the east.

China's most recent oil finds have been offshore. Surveying and exploration have centered on the Bohai area, east of Tianjin, which government officials have announced may hold more than 9.7 billion tons of oil and gas.[86] Phillips Petroleum announced in March 2000 that it had completed its appraisal drilling of the Peng Lai Find in Block 11/05 and would proceed with development. Full-scale production at the field is expected to reach more than 100,000 bbl per day by 2004. The Chinese company, Ocean Petroleum, meanwhile announced a plan to drill up to 80 wells in 2002 in the Bohai. Foreign partners in this plan include Shell, ChevronTexaco, Petronas Carigali, and Ultra Petroleum. [87]

Another major offshore field has been developed recently in the Pearl River Mouth area by a consortium including Chevron, Texaco, Agip, and CNOOC. The field began production in February 1999 and is expected to reach production of 27,000 bbl per day when fully operational. Meanwhile, improvement in Sino-Vietnamese relations is expected to open the way for oil and gas exploration in the Beibu Gulf (the northwestern arm of the South China Sea), especially since the countries have reached a preliminary agreement on a line of demarcation.[88]

China encourages foreign investment in exploration and infrastructure development.[89] CNSP is also pursuing several exploration ventures with foreign companies. By late 1996, almost 1 million square miles were open to foreign companies, including the southern provinces, the southeast sector of the Tarim Basin, and areas in northern and eastern China. Thirty petroleum contracts worth $770 million had been signed with 35 companies.

In July 2002, Premier Zhu Rongji announced that Beijing was opening the 2,500-mile-long West-East Pipeline project to international companies, specifically mentioning Royal Dutch–Shell, Gazprom, and ExxonMobil.[90] This pipeline would be a very significant national project, running from the Tarim Basin in Xinjiang Province through Gansu, Ningxia, Shaanxi, Shanxi, Henan, Anhui, Jiangsu, and Zhejiang Provinces to reach Shanghai. Its social and environmental impact is being studied by a joint UNDP–Shell project.[91]

There are two possible difficulties with this project, apart from funding and actually building it, which appear to reflect a lack of coherent planning at the national level. The first is whether there is sufficient natural gas in the Tarim fields to make the pipeline profitable; three major oilfields in the province reportedly produced 20.3 million tons (144.1 million barrels) in 2002, a threefold increase since 1990, but the evidence of proven natural gas is more problematic. The second reason for concern is the possible lack of planning at the national level to deconflict natural gas piped from the far west with that being imported from Indonesia, Papua New Guinea, and Australia.[92]

Russia's Far East is often cited as a potential major source of Chinese crude oil and natural gas imports, although so far there has been more talk than action. The Russian and Chinese governments have been holding regular discussions on the feasibility of pipelines to make such exports possible, with their first agreement signed in September 2001 to explore and develop jointly east Siberia petroleum reserves, estimates of which run as high as 11.5 billion tons.[93] In December 2002, Russian President Vladimir Putin and Chinese President Jiang Zemin issued a post-summit meeting statement that emphasized "the great significance of energy cooperation" between their countries, since "it is critical to make sure that Sino-Russian cooperation projects concerning crude oil and natural gas pipelines" are "carried out according to schedule" and that "the implementation of promising energy projects" be coordinated.[94]

Another project operated by East Siberian Oil at Yurubchenskoye is currently exporting crude oil to China by a combination of road and rail transport. Quantities are small, but East Siberian claims the field could have a production potential of 400,000 bbl per day if fully developed. Downstream infrastructure development in China centers primarily on upgrading existing refineries rather than building new ones, due to current overcapacity.

The petroleum refinery sector in China is overbuilt, although it suffers from a serious lack of adequate capacity suitable for heavier crude, which will have to be remedied as Chinese import demand rises, especially for Middle Eastern crude. It is hampered especially by a plethora of small, unprofitable refineries, although not nearly to the same extent that the coal industry counts too many small mining operations. Beijing has determined to shut down many of these small refining plants, although it also is concerned about resulting unemployment.[95] The problem of efficiency applies to larger production facilities as well, however: most Chinese refineries operate at financial losses that are hidden by state ownership. Chinese firms probably average $1.50 to produce a barrel of oil; Western companies typically spend $1.20. Natural gas exploration costs are similarly high for Chinese companies, $3.90 per barrel, compared to $3.00 for Western firms.[96]

Chapter Four

The Natural Gas Industry

Historically, natural gas has not been a major fuel in China, but it offers a very attractive alternative to coal and could relieve reliance on imported oil. Gas currently accounts for less than 3 percent of total energy usage in China, compared to a world average of 24 percent and an Asia-wide average of 8.8 percent. Beijing is trying to boost its production and consumption, but government officials complain that the sector is hampered by the lack of an effective regulatory framework.[97]

Beijing's goal for gas is 8 to 10 percent of the nation's total energy consumption by 2020.[98] This will involve an increase in domestic production and imports, either by pipeline or in the form of liquefied natural gas (LNG). LNG is natural gas that has been cooled to approximately 260 degrees Fahrenheit. LNG has a 610:1 volumetric advantage over its natural state and is much easier to ship and store. The import issue is complicated by the fact that in its natural form, gas can be piped only a limited distance without being liquefied. Furthermore, for liquefaction to be economical, the gas deposit must be 3 to 5 tcm in size.[99]

Natural gas is being sought both from its own reserves and in the form of liquefied natural gas from offshore and overseas sources. It is cheaper than either coal or oil and will ease the nation's reliance on those fuels while protecting the environment—natural gas is a relatively clean-burning fuel. If the suspected large reserves of natural gas are discovered in western and northwestern China, its use will ease reliance on foreign sources of energy.[100] Current estimates, however, project that by 2005, natural gas demands will exceed domestic supplies, and China will have to import the fuel.[101]

Given Beijing's domestic reserves and the environmental benefits of using gas, China has embarked on a major expansion of its gas infrastructure. The largest onshore source of natural gas comes from the

Changqing Oilfield in the Ordos Basin, which extends through almost 400,000 square kilometers (km²) of Shaanxi, Gansu, and Shanxi provinces, as well as the Inner Mongolia and Ningxia Hui Autonomous Regions. Two pipelines from this vast field—estimated to offer reserves of more than one tcm—currently transport gas to 15 cities, including Beijing.[102]

Total gas reserves onshore are estimated at 38 tcm, with proved reserves both on- and offshore of 2.56 tcm in 2000, a 58 percent increase over 1990 estimated reserves.[103] These are located in 69 different areas, the largest reserves in western and northwestern China, with major maritime sources in the East and northern South China Seas and in the Bohai.[104]

China would like to pursue what one observer has called the Pan-Asian Continental Oil Bridge, a network of oil and natural gas pipelines stretching from the Middle East, Southwest and Central Asia, Russia, and Southeast Asia to China.[105] However, the high degree of international political cooperation necessary for such a hyper-project is not likely to prevail in the foreseeable future; some proposals envision a grid extending from Southwest Asia to Japan, and from Indonesia to Russia.[106]

Even the western location of the major Chinese fields will require vast investments in pipeline infrastructure to convey the natural gas to eastern cities, which will remain the major market, despite Beijing's continuing efforts to direct future economic development to western provinces.[107] China is planning to build a pipeline 4,200 km long from gas deposits in western Xinjiang Province to Shanghai.

CNPC is seeking foreign investment in the project, with an estimated cost of $4.8 to $7 billion. ExxonMobil, BP, Shell, and the Russian company Gazprom reportedly bid for a stake in the pipeline, which is supposed to be in operation, with a transmission capacity of 12 bcm of gas, by 2003 or 2004. BP withdrew from the bidding process, perhaps because of a "lack of confidence in adequate financial return."[108] Several Japanese firms withdrew from the project for similar reasons.[109]

The contract has been awarded to a consortium composed of PetroChina, Shell, and Gazprom, with the two foreign companies holding 45 percent of the project, estimated at $18 billion, which probably includes development of the gasfields, construction of the pipeline, and development of associated infrastructure, including marketing.[110] Construction of the first section of the pipeline, 900 km in Shaanxi Province, supposedly began in August 2002.[111] Pipeline segments are also reported under construction in Zhejiang Province.[112] The Shell bid reportedly also includes the opportunity to operate one of the major fields, which would be a significant expansion of foreign involvement in China's energy infrastructure.[113]

The hope in Beijing is that these western fields will be able largely to support all the needs of China's eastern economy.[114] The pipeline is being touted under the three-part motto of "substitute oil with gas," "generate electricity with gas," and "city gasification."[115] At a more prosaic level, the Shanghai city government is planning to allocate this natural gas among electricity generation (50 percent), civil usage (40 percent), and for chemical industrial products (10 percent).[116]

This ambitious pipeline project raises socio-political questions similar to those associated with the mammoth Three Gorges Dam. Beijing apparently believes that developing energy fields in the barren, economically disadvantaged northwestern provinces will enable their populations to benefit from China's growing economy. That theory may be undercut by the historic fact that areas used as suppliers of raw materials rarely benefit significantly from having their wealth removed; more likely is that the economic benefits from new energy fields in the northwest will accrue to China's economically modernized eastern provinces. This will do little either to better balance the nation's wealth, or to relax significantly the uneasy socio-political-economic situation in Xinjiang and possibly other economically disadvantaged provinces.

Another proposed project would link the Russian gas grid in Siberia to China and possibly South Korea via a pipeline from Irkutsk. The cost of the project has been estimated at $10 billion, and a feasibility study is ongoing. Russian participation in pipeline construction in China will reportedly make Gazprom "Russia's largest trade and economic establishment in the Chinese capital," where it has opened a "permanent office."[117]

Expansion of China's gas infrastructure has the potential to create huge opportunities for foreign investment, as partners are being sought for both upstream and downstream gas projects. Enron and CNPC were planning to link Chongqing to Wuhan in central China with a 470-mile pipeline that would have an annual capacity of 3 bcm.[118] This pipeline was scheduled for completion in late 2002 but, despite Sichuan Province's large, proven gas reserves, is in jeopardy as the result of the collapse of the Enron Corporation, which held a 45 percent share of the pipeline.[119]

Shell is undertaking a study on developing the infrastructure to market production from the Changbei gasfield in the northwestern province of Shanxi, following a contract with CNPC to develop the field jointly, with estimated reserves of 2.5 tcf. BP Amoco's decision to purchase

20 percent of the stock offered in the PetroChina IPO has been seen as part of a larger strategic alliance with CNPC, which will likely lead in more substantial BP Amoco investments in China. BP Amoco controls the Kovykta gasfield in Siberia, the output of which could be exported to China, through its interest in the Russian firm Sidanco.

Other major natural gas projects in the offing include a northwest China to Shandong Province pipeline project,[120] and a huge LNG project in Guangdong. Imported LNG will be used primarily in coastal areas, and the Guangdong project is scheduled for completion in 2005, with an eventual capacity of 5 tcm per year; associated with this undertaking is construction of six 320-megawatt gas-fired power plants, and conversion of existing oil-fired plants with a capacity of 1.8 gigawatts to LNG. A feasibility study of a similar project in Fujian is under way.[121]

The Guangdong project originally drew interest from companies in Australia, Indonesia, Iran, Malaysia, Qatar, Russia, and Yemen, while investments in the gasfields in question have already been made by Dutch Shell, French TotalFina, ExxonMobil, and Australia LNG. In April 2002, Beijing narrowed the competition to companies from Australia, Indonesia, and Qatar.[122] China's goal is to obtain 3 million tons of LNG annually by 2005, and 5 million tons by 2008, for the Guangdong operation.[123]

Australia was awarded this contract to supply 3 million tons of LNG annually for 25 years to Guangdong Province.[124] This $13.5 billion arrangement will also allow CNOOC to "develop natural gas in Australia" as partner in a joint venture with Australia Natural Gas.[125]

Coastal China also will benefit from offshore production increases, such as the Yacheng field, which began production in 1996 and supplies 0.1 tcf per year for use in south China and Hong Kong. In January 2000, China granted preliminary approval to a project for an LNG regasification terminal at Shenzhen in Guangdong Province. Foreign firms will hold only a minority stake in the project.

Forecasts of significantly increased availability of offshore LNG assets are based on CNOOC determination to focus on natural gas as its "core business," as well as on three additional projects.[126] The first of these is a plan to convey 2.4 bcm of natural gas annually from two fields in the northern South China Sea, Dongfang and Ledong, to Hainan Island.[127]

Second is the annual transfer of 400 million cubic meters from the Bonan field in the Bohai to Shandong Province, while third is an ambitious plan to ship 3 to 4 bcm of natural gas from the Chunxiao and Xihu fields in the East China Sea to Shanghai and Ningbo. This last project is currently on hold, as the China companies in the proposed joint venture, CNOOC and Sinopec, try to reach a satisfactory agreement with their foreign partners, Shell and Unocal.[128] Shanghai, meanwhile, is pressing ahead with plans for the extensive infrastructure that will be required to focus its energy source on natural gas, infrastructure that will include the power-generation industry.[129]

An interesting aspect of Beijing's drive to increase natural gas supplies is an apparent effort, akin to that being pursued with respect to oil, to encourage its major energy companies to secure resources located in foreign countries. CNOOC has signed agreements with Chevron and Australia LNG to establish a joint venture to exploit fields in Western Australia. The CNOOC goal is to arrange for the conveyance of liquid natural gas to China from Australia's Northwest Shelf Gas Project, which involves BP, Shell, Chevron, and three other foreign companies.

CNOOC reportedly is also investigating investments in LNG resources in Iran, Malaysia, Qatar, Yemen, and Russia. A major investment has also been made in Indonesia's energy infrastructure.[130] Most recently, Beijing and Tehran signed a series of agreements that grant Sinopec the right to explore for oil in a 4,700-km^2 area south of Tehran, to upgrade Chinese-built refineries in Iran, and to build an oil terminal port on the Caspian Sea.[131]

Despite its drive to secure overseas supplies of oil, and significant experimentation with alternative sources of energy, Beijing has chosen natural gas as the future path to reducing dependence on coal, with its environmental and economic disadvantages, and on oil, with its strategic drawback of increasing reliance on foreign sources. Furthermore, locating natural gas and developing the infrastructure necessary to take advantage of it as an energy mainstay is emerging as a leading vehicle for foreign investment and opening of a core feature of China's expanding economy.

Chapter Five

Other Energy Sources

C hina has the world's fastest growing electric power industry, although the Ministry of Electric Power estimates that about 15 to 20 percent of the country's electrical demand is not being satisfied, with up to 100 million people still without access to electricity. China's 1995 electric generating capacity was estimated at about 190 gigawatts, about 75 percent of it produced from coal. Beijing has established a plan to increase the availability of electrical power, focusing on establishing major channels in the southern, middle, and northern parts of the country.[132]

Hydroelectric Power

China has the world's most abundant hydroelectric generating capacity, and it is a particularly important source of electric power in the central and western regions. But the location of this potential relative to markets and the environmental concerns associated with large projects could limit hydropower's contribution to China's electric generation needs. Nonetheless, the government seems intent on exploiting hydropower reserves "to obtain cheap, renewable, and clean energy."[133]

The Three Gorges Dam project on the Yangtze River, launched in 1993, involves construction of the world's largest hydropower project, with its 26,750-megawatt (MW) generating units slated to provide a total of 18.2 million kilowatt (kW) generating capacity by 2009, with an annual generation of 84.7 billion kW.[134] In October 2002, the Three Gorges Hydropower Station, announced as "the world's largest," was "formally founded" in Hubei Province, with an announced operating date of October 2003.[135] This project may cost as much as $25 billion, but even that figure may lack meaning, given the corruption, social cost of displacing over one million people and over 1,000 archeological sites, and unknown environmental aftereffects.[136] Additionally, this mammoth project's construction better represents China's inefficiency at distributing already available energy than it does an absolute need for new sources.

Nuclear Power

Beijing began construction of its first nuclear power generating plant in 1983, but 20 years later, nuclear power still represents a relatively minor share of China's electric generating capacity, with 6 plants currently in operation. Two are equipped with Canadian-built reactors (288 and 650 MW), at Qinshan at Hangzhou Bay in Zhejiang Province.[137]

A huge (1,812-MW) plant was also built at Daya Bay in Guangdong Province. Under construction are a second 650-MW unit at the Qinshan plant and two 1,000-MW units at a new plant, Lingao, near Hong Kong. The first two Qinshan reactors came on line in 1996 and 2002; the third is expected to begin operating in 2003.[138] Another large complex under construction is the four-generator, billion-kW nuclear power station at Shenzhen. The first of these began operation in late 2001.[139]

The Tenth 5-Year Plan (2001–2005) includes the construction of "some number" of nuclear power plants, with Shandong, Zhejiang, and Guangdong provincial officials taking the lead in planning new plants.[140] The government has not approved construction of a new nuclear power plant for 5 years, however, indicating the marginal regard in Beijing for this energy source.[141]

A senior nuclear power official has announced plans for 9 total units, which will produce 8.7 gigawatts (GW) by 2005. By 2015, output from nuclear plants is projected to increase ninefold over 1996 levels, accounting for about 4.5 percent of China's electric power generation.[142] Not addressed by this official, but reported in the press, is the nuclear power industry's difficulty in earning enough revenue to repay construction loans; about 90 percent of the money for the Qinshan project was borrowed from Canada, the United States, and Japan.[143]

A new plant near Beijing reportedly incorporates state-of-the-art "high-temperature air-cooled" technology. Its status as "one of the 15 key projects of the '863' state high technology research development program" indicates that it may be linked to the military sector.[144]

Chinese plans for increasing nuclear power generating capacity usually mention Russian assistance, but the actual degree of collaboration is uncertain. Although Moscow is providing various equipment for the construction of the Tianwan nuclear power plant, including manufacture and installation of the first complete reactor, actual plant construction has not matched the rhetoric.[145] Overall, this is a modest plan, and Beijing is planning a limited role for nuclear power in satisfying China's energy needs, probably no more than 4 to 5 percent by 2020.[146] One reason for this is the lack of indigenous capability in nuclear power plant construction.[147]

Renewable Energy Sources

China's renewable energy resources include biomass (garbage), ethanol, geothermal, solar, and wind. Further development of these resources could reduce China's growing dependence on imported oil and its need for additional coal-fired power plants and provide sources of energy for populations in remote areas not currently served by existing energy distribution systems. None of these sources, however, should be expected to make more than a slight dent in China's energy needs.

Investment in development of wind energy resources is expanding, and the government has announced tax breaks for wind power generation producers. As of 1995, the country had 44 MW of wind power generating capacity, of which 14 MW were installed in 1995 alone. Xinjiang Province, in northwestern China, has announced plans to add 66 wind power generators to an existing plant to create what would be the largest wind power base in Asia. Fujian Province has announced plans to build wind generation power plants with a capacity of 200,000 kW by 2005.[148] Guangdong Province is also pursuing development of wind power plants.[149] The nation's total wind generating capacity was approximately 375 MW in 2000, and could exceed 1 GW by 2010.[150]

Beijing is also investing in development of its geothermal energy resources, with nearly 29 MW of generating capacity developed as of 1995 (up about 50 percent from 1990). As with wind energy, the market potential and resource base are significantly greater: an estimated 1,800 MW of geothermal resources, with market potential of 600 MW.

China's solar energy resources are assessed at 4 MW per square meter, with market potential for solar energy estimated at 135 peak MW. Biomass resources are assessed at 260 million tons of oil equivalent.[151] International assistance for developing China's renewable energy resources includes support from the World Bank; since 1998, the bank's Global Environmental Fund has been funding appraisal projects for photovoltaics (conversion of light energy into electricity), wind, and biomass (conversion of refuse into electricity), as well as hydroelectricity. Finally, a program promoting ethanol-based fuel has been launched, both to benefit the environment and to improve the grain market for farmers.[152] A three-test site program for processing landfill gas is also being funded by the national government.[153]

Chapter Six

The Power Distribution System

Building a national power grid is certainly the most ambitious—and probably the most important—feature of the Tenth 5-Year Energy Plan. Under the Ninth (1996–2000) Plan, China intended to add approximately 16 GW of generating capacity annually (including about 3.5 GW per year of hydropower). The goal was to achieve 290 to 300 GW of installed capacity by 2000. China also planned to expand its electric power transmission system, link existing grids, and implement a unified national power grid by 2020. Reportedly, over $18 billion was invested in 1999–2001 upgrading rural power grids.[154]

Beijing now has stated that it will commit over $43 billion to grid construction over the span of the Tenth 5-Year Energy Plan (2001–2005), while the National Electric Power Corporation (NEPC) is planning to spend $72.5 billion on power distribution projects during that same period.[155] Improving the supply of electricity to rural areas is particularly important to the national government as part of the plan to improve economic conditions in economically disadvantaged regions.[156]

The national energy investment figures for the first half of 2002 indicate that the government is serious about modernizing the power grid system, since that sector received the largest share of new investment funds.[157] This emphasis recognizes the current reality in China's energy sector: the nation suffers not from a shortage of electric power generating capacity, but from an inadequate distribution grid for that power.

The issue facing the state is how to distribute the available supply of electricity more efficiently, either by improving distribution or by increasing the overall supply of power. The first solution may be the most efficient in economic terms, but the second course of action may more quickly achieve the desired result. Beijing is compromising, pursuing both grid improvement and supply increase.

This is truly a national scheme, with power plant construction planned for the cities of Beijing, Tianjin, Tangshan, and Shanghai; for Jiangsu, Zhejiang, Hebei, Hunan, Guizhou, Guangdong, Guangxi, Yunnan, and Gansu Provinces; and for the Tibet, Qinghai, and Inner Mongolia Autonomous Regions. Plans for Guangxi include an 8.6 billion yuan investment in four power plants, Gansu has received central government approval for four power projects, while the four hydropower plants planned for Yunnan may eventually exceed the generating capacity of the Three Gorges system. This would be yet another massive project in China's southwest, with the coming on line of the second of eight hydropower plants planned for the Lancang (Mekong) River.[158]

One expert has concluded that China will need to spend about $100 billion to turn its "very fragmented" power distribution system into a linked, well-regulated system capable of providing even regional distribution of power throughout the country.[159] As is the case with expansion of petroleum sources, China is intent on attracting foreign investment to finance this cost. The government aims to "improve the investment environment for the power industry" as part of the effort to attract large-scale foreign investment, with a goal of attracting 20 percent of the funding for the electric power sector from foreign investment. This goal was formalized in China's first law governing electric power generation, which was enacted in 1996. In its first project open to international bidding, China awarded a build-operate-transfer project (the 720-MW coal-fired plant in Laibin, Guangxi Province) to a consortium headed by France's EDF. Bidding on a second project (a 600-MW plant in Hunan province) is under way.

The Tenth 5-Year Plan continues to pursue this objective, with the State Development Planning Commission submitting a plan in October 2002 for approval by the State Council. This proposal includes creation of a National Power Regulatory Committee, two power grid corporations, and five independent power generation groups, each with responsibility for a specific region. These regional bodies will implement state power policy and oversee the trans-power market.

The State Power Corporation, which owns half of China's installed electricity capacity and its entire transmission network, has been broken up into five generating companies: the China Huaneng Group, China Datang Group, China Huadian Group, China Guodian Group, and China Power Investment Group. Two grid operators, the State Power Grid and the China South Power Grid, will oversee the groups in north and south China, respectively. The State Power Regulatory Commission will oversee the new organizations.[160] Beijing has also announced it will devote $45 billion, 40 percent of the Tenth 5-Year Plan's investment in power projects, to expanding the nation's power grid.[161] Zhang Guobao, vice minister in charge of the State Development Planning Commission, officially announced this new break-up strategy in Beijing's Great Hall of the People in December 2002.[162]

Production and distribution of energy will remain one of China's greatest challenges in coming years. Too many energy reserves are far from consumption centers, and bottlenecks exist in transportation and electricity distribution. Additionally, industrial concerns produce significant amounts of electricity for which only limited statistics are available. In energy-intensive process industries such as chemical and steel production, refineries, and mining/minerals, plants typically generate their entire internal electric requirements and export excess electricity to the local community.

To meet its enormous appetite for electricity, China will need to access some 20 GW of additional generating capacity each year for the foreseeable future. China's insufficient power availability is exacerbated by underinvestment in transmission facilities. Inefficiencies and power losses have also resulted from a mismatch of generating capacity and transmission capacity, with the latter insufficient to service China's burgeoning industrial and residential demands.

Chapter Seven

The Environment and the Energy Sector

China's significant environmental problems are due largely to the lack of modern technologies and remedial policies needed to update the infrastructure. Air quality is a major health concern. The cities of northeastern China especially suffer from polluted air, but the problem has accompanied industrial development throughout the country: 9 of the world's 10 cities with the most polluted air are in China.[163]

Beijing announced the latest State Industry Technology Policy in June 2002, which, in conjunction with the Tenth 5-Year Plan, establishes as "major targets" the use of "clean energy technologies, as well as oil substitution."[164] Specific goals include reducing the power industry's sulfur dioxide emissions by up to 20 percent from 2000 levels, probably an unrealistically ambitious goal.[165] The country's environmental emission levels can be significantly lowered, however, simply by taking advantage of existing technology.

The question remains whether the national and provincial governments are able to find the resources to fund and to convince the power industry to make the changes necessary to meet this ambitious goal. The scope of the problem may be indicated by a report that "Beijing residents are expected to inhale 8,500 tons less of sulfur dioxide and 4,300 tons less of soot this [2002–2003] winter" due to such improvements as coal-burning boilers being "rebuilt."[166]

Existing environmental problems already pose very serious challenges. The ever-increasing energy demands generated by rising living standards and economic growth increase the urgency and scale of addressing the problem. The problem is being addressed, with the Special Olympics planned for Shanghai in 2007 and the Olympics scheduled for Beijing in 2008 serving as a spur.[167]

The SETC has announced a Green Project in support of the Olympic Games, emphasizing increased use of solar energy, improved water conservation, and greater reclamation of recyclable materials.[168] The city is also planning to use geothermal energy in Olympic facilities.[169] The scope of the city's concern is also indicated by efforts to invigorate the Green Light Program, begun in 1996 to attempt to reduce mercury contamination.[170]

The environmental consequences of continued heavy use of coal raise important issues not just for China but also for global efforts tackling the problem of climate change. Beijing has announced a goal of reducing its annual consumption of coal from 27 million tons in 2000 to less than 15 million tons in 2008. It has also set a goal of having "8,000 buses and 40,000 cabs . . . fueled by green energy" by 2007, which would account for 90 percent of the former and 70 percent of the latter. The primary means of accomplishing this ambitious standard will be through providing natural gas stations.

Other cities are following suit. A Shanghai official has claimed that the city "has made it a long-term target to build itself into an ecological metropolis."[171] If the massive West-East Pipeline is constructed, Shanghai's sulfur dioxide pollution might be reduced by 90 percent and the current acid rain problem largely resolved.[172] China is now the second largest emitter of greenhouse gases in the world (after the United States), while acid rain falls over 30 percent of China's landmass. The World Bank estimates these conditions to cause 178,000 premature deaths a year in China and in 1995 to have been responsible for as much as $13.75 billion in economic losses.[173]

As the world's largest producer and consumer of coal, China should be able to focus on managing and balancing the environmental consequences of its dependence on coal as the major fuel of its economy. Beneficial results are possible from instituting new technology and processing policies. These range from improved mining methods, which would reduce methane emissions, to better management of water and land resources.

For the present, however, the overwhelming dependence on coal for fuel continues to affect the environment severely. Many of China's environmental problems resulting from dependence on coal can be attacked by implementing existing technology in both mining and using coal. There is, however, strong competition in China for limited capital to fund the widespread modernization of the economy. Hence, Beijing faces difficult prioritization issues when allocating financial resources to the implementation of new technology for producing energy: does it devote available capital to spreading the application of current technology, or does it devote that capital to emerging technology, in an attempt to "leap ahead" of current shortfalls?

The linkage between non-hydrocarbon energy sources and the environment is illustrated in a report of "small, eco-friendly hydropower stations" being constructed. A tiny hydropower station built in Tibet reportedly allowed a farm family to substitute electric heat for the 20 kilograms of wood burned daily. This in turn should slow the deforestation process and thus preserve the habitat in which pandas flourish.[174]

Much progress has been made in China during the past quarter-century to lessen the severe environmental degradation that has accompanied the nation's dramatic economic development. During the past 5 years, moreover, the national and, to a lesser extent, provincial governments have begun instituting policies to counter environmental problems.

Government action has been particularly and increasingly pressing in the case of fresh water, with $3.6 billion earmarked for improving the water supply.[175] Especially in northern China, historic shortages have been exacerbated by drought, population increases, and economic development. The demand for water for personal and business uses has outstripped the supply. One analyst has claimed that China has just 8 percent of the world's fresh water supply to support 22 percent of the world's population.[176] Water also has very serious implications for Asia's international relationships, as China's efforts to assure its own future inevitably will clash with those of other regional nations.

Thirty million urban citizens were recently estimated to lack adequate fresh water, including those living in 400 of China's 668 cities.[177] Beijing's response to this increasingly critical shortfall has been multifaceted, urging conservation on the part of consumers, proposing a nuclear-powered desalination plan, and, more meaningfully, launching a massive project to divert water from southwestern China to the northern part of the country. This project will link the Yangtze and Huaihe Rivers and the Yellow River and Haihe valleys, forming a network of four horizontal and three vertical waterways that allows a more productive allocation of the country's water resources nationwide.

This ambitious plan is designed to divert 3.8 trillion cubic meters of river water a year, which equals the annual flow of the Yellow River. This project was initiated in 2001, with worldwide public bidding invited for specific phases. While details are lacking—indeed, planning is still under way—the first and second of this plan's three routes are scheduled to be under construction by 2010. Total cost will probably exceed $50 billion.[178]

This water diversion scheme rivals the Three Gorges Dam project as an engineering feat and will probably surpass it in political importance. First, there are bound to be domestic implications, as water resources are taken from one province and given to another. Second, the nations of South and Southeast Asia have already begun registering their concern about Beijing following diversion of the Yangtze River with attempts further to "harness" the other three great rivers with headwaters in the same area.

The Mekong, Irrawady, and Indus Rivers are of the utmost importance to the economic and social existence of the nations lying downstream; some of these have begun complaining that China's riverine efforts have already reduced the flow of water down river.[179] In the meantime, Beijing has promised to cooperate with the Mekong River Commission to ensure equitable distribution of water, environmental preservation, and electrical power sharing.[180] It is providing hydrological data on the river's upper reaches and in June 2002 hosted the latest meeting of The Great Mekong Subregion Cooperation group, which was organized in 1992. In addition to China, Burma, Cambodia, Laos, Thailand, and Vietnam sent representatives, as did the Asian Development Bank.[181]

The quality and sufficiency of water for China's continued economic growth is far more than an environmental question; it has a direct influence on continued social coherence among that nation's vast population, whose growing sense of nationalism and loyalty to the Beijing regime depends in large part on continued economic growth at a personal as well as a societal level.

Implementation of new technology may be slowed by China's desire to emphasize indigenous development. The continued rapid pace of economic development and its deleterious environmental effects mitigate against such an autarchic policy—the current state of air and water pollution and shortages means that Beijing must pursue solutions whatever their origin, domestic or foreign. Beijing recognizes the need for economic and technical cooperation between foreign and domestic organizations but insists on doing so according to its own terms and schedule. Nonetheless, it is doubtful that any significant energy infrastructure project has been launched during the past decade without very significant foreign involvement. One proposed project, the "main body of the Zhiganglaka

Hydropower Station on the upper Huang He," will be "the first state-listed power project exclusively funded by overseas investment," by the American AES Corporation and a Hong Kong company.[182]

The growing degree of foreign involvement in China's energy sector increases Beijing's concern with the global energy situation, expressed at the November 2000 meeting of the Asia-Pacific Economic Cooperation (APEC) group by Foreign Minister Tang Jiaxuan. Alternate sources of energy are being investigated, both for generating electricity and powering vehicles; they include generating power from rubbish, straw, and marsh gas processing; wind power; hydropower; hydrogen; and solar energy.[183] A primary concern is the degree to which alternate sources are less injurious to the environment.

China seems determined to improve its record of conserving and improving the environment. It must do so not only for the near-term objective of the 2008 Olympic Games but also for long-term economic and social reasons. At a September 2000 APEC workshop on offshore energy facilities and the environment conducted in Beijing, the Chinese representatives emphasized the importance of environmental protection along the country's very long (18,000 km) coastline and noted the revised body of law on protecting the marine environment China has instituted.[184] Beijing is advocating corrective policies and pushing the requisite programs, but the struggle to gain the effective support and cooperation of Chinese society, especially its industrial sector, is far from won.

Chapter Eight

Energy in the Ninth and Tenth 5-Year Plans

Beijing, as part of the Ninth 5-Year Plan, aimed to increase total energy output approximately 9 percent by 2000. This plan's success involved the accomplishment of supporting programs. The first was to improve energy efficiency by 5 percent annually, a goal requiring significant technology improvements in 15 industries, including coal and electric power industries, and improving the efficiency of the iron and steel, nonferrous metal, chemical, building material, and transportation industries. The second major objective of the Ninth 5-Year Plan was to increase the conservation of resources, described as a "top priority" but qualified by the equal priority given to "development."

The plan's focus in power development was on coal, oil, and gas exploration but also emphasized developing new energy sources. Beijing discussed at length the availability of electric power as a final result of the plan. The goal in this vital sector was to increase capacity and generation by 7 percent annually, to reach 290 GW in generating capacity and generate 1.4 trillion kW-hours annually by 2000. This in turn included the simultaneous promotion of hydro and thermal sources; "appropriate" development of nuclear power; the development of power stations near coal mines with a high-parametric, high-efficiency capacity of at least 300 MW; and an emphasis on flue gas desulfurization and extra-high voltage transmission technologies.[185] Other new generation sources, including wind, marine, and geothermal power, were to be developed.

In the critical area of coal production and utilization, the Ninth 5-Year Plan aimed to increase total output to 1.4 billion tons by 2000, with the emphasis on stabilizing output in the east and developing mines in Shanxi, Shaanxi, and Inner Mongolia. Other steps included the accelerated development of technology for cleaning coal and using high-quality anthracite from Shanxi's Jicheng as a base for chemical fertilizers.

For oil and natural gas, Beijing aimed by 2000 to boost proven reserves by 33 billion bbl of crude oil and 17.7 tcf of natural gas; to increase crude oil output to 3.1 million bbl per day and refinery output to 4.5 million bbl day; and to increase natural gas production to 833 bcf. These increases were to lead to the conversion of 70 percent of urban households to gas fuel.

The Ninth 5-Year Plan recognized that some output would have to continue to come from overseas resources. The plan emphasized the importance of geographic balance, the first element of which was called the *onshore principle*, which meant stabilizing production in the east while increasing production in the west. Oil and gas were to be exploited simultaneously and their scope of development expanded.

The second element of geographic balance was called the *offshore principle*, which meant continuing to open and expand the scope of operations to exploit offshore gas and oil. The hope was for steadily increased production, while also stepping up efforts to discover additional resources and increase proven reserves.

Beijing hoped to maximize the use of natural gas from Hainan Island and from Xinjiang Province, increase the use of oilfield products for large-scale nitrogenous fertilizer plants, and upgrade and expand refineries in Zhenhai, Maoming, and Fujian. Environmental concerns were addressed in the goals of ending production of leaded gasoline and adding and improving pipelines and storage facilities.

How successful was this plan? According to the U.S. Department of Energy, China's energy situation in late 2000, based on annual consumption of 36 quadrillion btu (quads), included 887 billion kW of electricity generated (approximately 70 percent coal-fired); 0.7 tcf of natural gas produced (representing about 2 percent of energy utilized); 3.6 million bbl of oil consumed daily (an increase of more than 80 percent since 1986); and coal consumption of 1.5 billion tons (fulfilling 75 percent of the nation's energy needs).

The oil sector has apparently led the field in identifying additional resources during the Ninth 5-Year Plan. In 2001, Chinese companies drilled more than 920 wells, with 48 percent located in the already-well-established Daqing and Shengli fields. Offshore production accounted for about 10 percent of China's total output, with about 60 new wells drilled (60 percent by foreign oil firms).[186]

The Ninth 5-Year Energy Plan, then, while failing to achieve all of its goals, marked a significant expansion of China's energy sector, in terms of both domestic and foreign sectors. Beijing was able to support the continued economic growth it deems necessary to satisfy China's people and maintain its hold on power.

The Tenth 5-Year Plan

China will be focusing on several areas in the Tenth 5-Year Energy Plan (2001–2005). Ensuring access to reliable energy resources that are plentiful enough to support continued economic growth is vital. This concern is being addressed by:

- expanding acquisition of energy resources around the world, as well as further developing sources in Chinese territory ashore and at sea. The policy is susceptible to changing international developments, including the downfall of tenuous governments. Even more problematic may be China's dependence for over half of its imported oil on Southwest Asia and the Middle East, areas of doubtful long-term stability.

- taking advantage of foreign capital and technology to discover and exploit energy resources by adopting foreign policies amenable to securing the resources, such as those in Central Asia. This strategy is required for future expansion of the energy sector but also will constrain Beijing's foreign policies in proportion to its reliance on foreign-controlled energy sources.[187]

- conserving and reducing environmental pollution, which is inherently wasteful of energy resources. This program will have to be more carefully and constructively managed than it has been in the past if China is to reduce the pollution attendant upon economic growth.

- pushing Chinese industries to adopt energy-conserving technologies and urging development of high-tech industries that require relatively little energy. Here, Beijing will have to continue relying on technology of foreign origin to buttress indigenous efforts at privatization and modernization of the industrial sector. This dependence will influence China's foreign policy choices.

- establishing a national strategic oil reserve equal to at least one month's supply.[188]

These broad goals are based on the assumption that China's dependence on foreign energy sources will continue to increase. The last step is a product of concerns about a volatile world petroleum market, especially in view of current uncertainties in the Middle East and Southwest Asia.[189] The plan also includes an ambitious concept for upgrading China's petroleum industry, namely to:

- complete a network operational system in 3 to 5 years
- raise the success rate of exploration by 3 percent
- shorten the cycle of oil well drilling by one-third
- raise the oil recovery rate by 2 to 3 percent
- raise the proportion of oil and gas in the primary energy to over 3 percent
- raise the contribution by science and technology to oil and gas exploration to 55 percent.[190]

These objectives must be considered within the national political goal of better integrating China's eastern and western regions. One facet of this rubric for the energy sector is the West Electricity for East Program, which was officially initiated in June 2002 when a 380-km, 500-kilovolt power cable from Chongqing to Shanghai began transmitting power.[191]

Beijing thus will have to resolve, both for reasons of continued economic well-being and domestic political tranquility, a national priority of working toward a stable international environment. This is especially true in East Asia, where the predominantly maritime character of trade will increasingly determine the reliability of the energy supply central to China's continued economic health and societal satisfaction.

Another international issue is Japan's reliance on China as a source of coal and petroleum. For 2001, this dependence will exceed 3 to 4 million tons of crude oil from China's Daqing oilfield alone and more than 7.5 million tons of coal for each of the next 3 years.[192]

U.S. Department of Energy projections based on a 4 to 5 percent annual growth in energy demand for China (compared to about 1 percent growth in the world's industrialized countries) indicate that this dependence will continue increasing for at least the next 15 years.[193] This report describes three key facts about China's ability to provide its citizenry with energy products: fully one-fifth of China's population is still without electricity; energy production still suffers from very significant shortfalls in domestic transportation—crucial in view of the great distances of significant energy sources from population centers; and Beijing is still struggling to solve the problem of inadequate distribution of the energy already produced.

China's biggest challenge during the Tenth 5-Year Plan (and beyond) will be to diversify energy production from the current overwhelming dependence on coal. By 2015, the share of the non-coal sectors will certainly increase but not by enough to offset the overwhelming dependence on the essentially dirty energy produced from coal. This dependence is actually expected to grow slightly to 77 percent of China's overall energy production by 2015.[194]

China is expanding the infrastructure needed to support the steady, significant increase in energy resources required to sustain continued economic growth. Infrastructure shortcomings reflect energy resource problems not only with availability but also with distribution of available supplies. Hence, China is expanding rail, riverine, and ocean facilities for moving petroleum products.[195]

Perhaps most telling for future efficient use of energy to support economic growth and popular satisfaction is Beijing's announced plan to reorganize and expand China's power distribution system, both to increase the efficiency of power utilization and to "create and safeguard a sound market environment for fair competition."[196] Finally, Beijing is attempting to instill a sense of conservation in the exploitation and consumption of energy resources. This emphasis fits with policies to combat the environmental degradation that so seriously affects all sectors of Chinese industrial development. One government pronouncement declared "a lack of energy resources and the increasing environmental pollution have become important factors that reign in our economic and social development." [197]

Future Oil Imports

China's oil imports are increasing yearly, almost doubling during 2000 alone, with the majority coming from Middle Eastern countries and Southwest Asia.[198] During the first 9 months of 2001, for instance, Iran exported 63 million bbl of oil to China, more than any other nation. Saudi Arabia and Oman were the second-leading exporters to China, each providing 44.1 million bbl.[199] This trend will almost certainly continue, as China's population continues its nascent migration from rural to urban areas, a classic factor in a great nation's economic modernization.

China produced 50 percent more automobiles in 2002 than in 2001, but currently there are just 10 motor vehicles for every 1,000 Chinese citizens, while the comparable numbers are 30 for Egypt, 552 for Japan, and 770 for the United States. This disparity indicates how far

China's energy demands are likely to grow, as disposable income increases and World Trade Organization (WTO) membership results in lower automobile prices for the huge population.[200] Indeed, one observer has noted that *"per capita* oil use in China is nearly 30 times less than that of the United States"[201] and "from 1978 to 1995, the number of passenger vehicles per thousand people increased nearly 1,000 percent." Hence, he offers the "possibility of up to 250 million cars in China" by 2050, a particularly striking figure as an indicator of possible future energy demand in the nation in light of the fact that there were only 700 million cars worldwide in 1999.[202] The potential for China to become the "number one car market in the world" contributed to the government establishing the National Clean Automobile Taskforce in 1999, which recently announced a $120 million project to develop "environmentally friendlier motor vehicles by 2004."[203]

Chinese officials are concerned about this growing reliance on foreign sources of the energy crucial to China's continued economic health, especially in view of the recent but prolonged increase in the price of crude oil.[204] Nonetheless, China continues to rely on foreign capital, technology, and know-how to recover domestic energy resources, cooperating with more than 50 foreign companies to exploit just the offshore oil sector since 1979.[205]

The government apparently has decided on a multilevel approach to ensuring adequate fuel supplies. First is expanded domestic production, including both exploring for new energy fields ashore and at sea, and usually with foreign participation. CNOOC, for instance, has announced its consortium with U.S.-based Phillips Petroleum to invest in large-scale offshore oil exploration during the next 5 years.[206]

Second is a program for increasing the effectiveness of the process for turning the raw product to usable energy, including tight control of refineries.[207] Third, Beijing is signing contracts with a wide range of countries that will export oil to China. Fourth is the drive to gain exploration rights and production control over energy fields in foreign nations.

Fifth, Beijing is creating an energy reserve to provide a ready source of emergency resources and also "to avoid massive economic losses resulting from fluctuations on the international market." This last program includes steps to control the rate of increase in fuel consumption, specifically by "ensuring the stable growth of domestic crude oil production, accelerating the development and use of natural gas, and promoting the restructuring and technical upgrading of the petrochemical industry."[208] Instability in the Middle East[209] in part drives the creation of a national energy stockpile, which may be as large as 4 million cubic meters valued at $10 billion.[210]

The idea of a strategic petroleum reserve has gained urgency since the September 11 terrorist attacks in the United States: a senior State Council economist has noted that American military campaigns in Afghanistan and possible campaigns in Southwest Asia "will greatly affect China's oil strategy in the Middle East and Central Asian countries."[211] This program aims to reduce possible energy shocks by establishing an oil stockpile equivalent to at least 3 months' national supply.[212]

In an August 2001 statement, the State Economic and Trade Commission announced that China would establish a "state strategic oil stockpile . . . by 2005" as a buffer against "the volatile oil market." The plan reportedly calls for a 30-day supply of fuel oil and would cost more than $2.5 billion.[213] This program has also been endorsed by the State Development Planning Commission, PetroChina, and, interestingly, by the U.S. Department of Energy.[214]

China is also considering reestablishing a market in oil futures open to public investors, although this will likely not occur until 2005. [215] Two factors are probably causing the government to hesitate: WTO membership, which may enhance the desirability of a futures market, and the possibility that a futures exchange will require the government to change the way it currently manages and controls the oil market.[216]

The government has ruled that foreign companies are not limited in the number of shares they can buy in businesses connected directly to the proposed West-East Pipeline. Beijing allows foreign companies to invest even to the point of purchasing controlling shares in pipelines. This policy represents China's decision, at least in the energy sector, of allowing foreign control over major parts of the energy infrastructure on which the nation depends.[217]

The government also intends to continue controlling the import of refined petroleum products, trying to strike a balance between reliance on foreign sources and the health of the domestic economy.[218] This balance will be difficult to achieve, given Beijing's apparent decision to end the national oil monopoly by floating major companies, such as Sinopec. A further complication is that foreign oil companies can compete in the retail market, especially in refined oil products and the operation of automobile service stations.[219]

In fact, this sector of the energy economy has broader social and economic implications, since Beijing reportedly is trying to "rectify the gas station market" by closing illegal stations and forcing others into semi-monopolies controlled by PetroChina and Sinopec, which currently operate about half of the national total.[220] The government is urging these companies to form cooperative operations with Shell, Exxon, Mobil, and other multinational oil companies. British Petroleum and PetroChina have already announced such an agreement.[221] The prospect of rural drivers buying fuel from foreign stations apparently is palatable to Beijing, although it would no doubt be anathema to earlier generations of CCP leaders.

Entry into the WTO also raises many concerns for Beijing. At a November 2001 International Symposium on the Fuel Oil Market in Guangzhou, government officials emphasized that WTO entry "would inevitably bring heavy blows to" China's oil industry.[222] While Chinese energy industry analysts do not universally support this judgment, it has introduced an element of uncertainty into public and private sector attempts to develop the nation's energy infrastructure. Further, Beijing will have to tread a fine line between meeting WTO requirements and fostering the expansion of the energy sector required for continued national economic growth.

One step being taken is a significant reduction of the tariff on imported refined oil from the current 69 percent to 6 percent.[223] Import quotas for petroleum products have also been increased by 15 percent, in accordance with WTO conditions.[224] According to one writer, foreign oil giants "are scrambling to get their foot in the door [with] visions of a gas station on every corner in a country with 1.2 billion potential consumers."[225] It also hopes that WTO membership will provide avenues for influencing the price of petroleum.[226]

Beijing seems determined to participate fully in the WTO regime to maximize benefits to the economy, including the energy sector. In other words, as China drives to optimize the availability and efficiency of the resources it needs to fuel its energy industry and hence its economy, it is casting its net of programs widely and imaginatively.

Beijing aims to exploit foreign petroleum sources while avoiding reliance on foreign production. Instead of merely importing crude or refined petroleum from other countries, Beijing is pursuing agreements to obtain exploration and production rights for likely petroleum fields located in foreign states.[227]

China's desire to diversify its energy resources is indisputable. Beijing would like to expand significantly the role natural gas plays as an energy source and is investigating non-hydrocarbon energy sources, but it is having difficulty achieving a near-term, significant reduction coal consumption. Importantly, no clear sense of prioritization emerges from the plethora of initiatives, plans, and declarations emanating from Beijing. This apparent lack of prioritization applies across the board, with massive pipeline projects probably competing with other huge infrastructure projects, such as the Three Gorges Dam system or the plan to divert water from southwestern to northeastern China.

Chapter Nine

Defense of the Energy Sector

China's future political stability and national coherence depend in no small part on continued economic growth fueled by readily available, affordable energy supplies from both domestic and foreign sources. The People's Liberation Army (PLA) is deeply involved in the energy sector at two levels. First, the military frequently is tasked with participating in the modernization and expansion of China's energy infrastructure. The 2002 Defense White Paper issued by Beijing notes PLA participation in "the construction of nine energy facilities such as pipelines, natural gas fields and oil-and-gas fields; the construction of seven hydropower stations and nineteen trunk diversion channels [and for] the protection and construction of the ecological environment."[228]

The second level of PLA involvement relates to the question of possible threats to domestic sources. The White Paper notes that the military is responsible for "maintaining and promoting social stability and harmony," to include "cracking down on all criminal activities that threaten public order."[229] Beijing typically classifies as "threats to public order" incidents of labor unrest and other similar activities, which means that the People's Armed Police (PAP) and the PLA are primary instruments of government control as the energy infrastructure undergoes the sometimes traumatic effects of privatization and modernization. The military is especially concerned, of course, about possibly violent threats to power sources and pipelines.

The advent of the Three Gorges Dam project has drawn the attention of China's national security analysts. PLA planners supposedly have discounted for the immediate future the possibility of air raids on this massive installation because of its location in the middle of the country, China's nuclear deterrent, and the nation's air defense system.[230] Despite those reports, the People's Liberation Army Air Force must consider how to defend such an important piece of the national infrastructure.

Uighur separatists and other non-Han groups in China have long been a low-level but nonetheless bothersome factor in maintaining domestic tranquility. Most serious has been unrest in Xinjiang, the potentially energy-rich western province subject to transnational Islamic movements. Separatist threats have been highlighted since the events of September 11. The war on terrorism proclaimed by Washington has, de facto, focused on Islamist groups, a factor that probably provides the primary motivation for Beijing to join the United States in this struggle.

Categorizing such groups as terrorist legitimizes oppression and persecution that hitherto would have drawn condemnation from the United States and other Western nations. When the Deputy Secretary of State, Richard Armitage, visited China in August 2002, he supported classification of the Uighur group Eastern Turkestan Islamic Movement (ETIM) by the United Nations as a terrorist organization, much to Beijing's satisfaction.[231]

Some steps have been taken to protect domestic energy facilities against organized attack, although defensive measures appear limited to those by local police and PAP units and those attendant to routine defense plans of Military Region staffs. For instance, in mid-2001, the State Council promulgated revised Regulations for Protecting Oil and Natural Gas Pipelines, a measure intended to defend against "seizing, sabotaging, stealing, and looting pipelines and facilities."[232] In one recent case, "a gang" stealing power cables to sell for profit was prosecuted.[233]

More serious events affecting the production of domestic energy supplies have resulted from labor unrest, primarily in the Daqing petroleum complex, which began producing in 1960 and has long been China's most productive field. Reorganization of the Daqing operation, site of China's largest and most productive oilfield, reportedly has resulted in the loss of approximately 80,000 jobs.

One source has estimated that the unemployment rate in the rust belt provinces may be as high as 25 percent, which has contributed to "workers rioting over job losses . . . challenging the police and demanding social security benefits."[234] As a result, up to 50,000 laid-off workers and their families demonstrated throughout March and April 2002.[235] As many as "several hundred thousand workers from the Daqing Oilfield" may have participated in various demonstrations.[236]

The economic problems in China's northeastern rust belt provinces have been exacerbated by the government's determination to overhaul inefficient SOEs. It has involved the displacement of long-term workers, often without the compensation promised during their employment. Local police, PAP, and regular PLA troops were all used to counter resulting labor demonstrations during 2001 and 2002 that frequently involved violence by both sides.

Similar incidents have occurred elsewhere in China's vast energy infrastructure.[237] Reports are sketchy, given Beijing's understandable sensitivity to such unrest, but a particularly violent conflict was reported in Shanxi Province, where 2,000 coal miners at Datong went on strike.[238] Other major incidents have been reported, one involving 20,000 miners in Laioning Province, and a second involving thousands in Guizhou Province. Indeed, one knowledgeable Western correspondent believes that incidents of large-scale civil unrest occur almost daily in China, with the oil industry particularly susceptible as a result of information exchanges established between workers from different and widely spread fields.[239]

The privatization of SOEs, among which the energy sector has been prominent, exemplifies one of the most difficult aspects of modernizing China's economy: the need to maintain societal peace by providing a hitherto absent social safety net. Ironically, the transformation of the economy from communism to capitalism has required China's leadership to address at the most basic level the societal inequality that has historically precipitated revolutionary developments in industrializing societies.

At the beginning of the 16th Party Congress in November 2002, President Jiang Zemin included in his eight national priorities the need to "deepen the reform of the income distribution system and improve the social security system," and to "do everything possible to create more jobs and improve the people's lives." He also said that "developing socialist democracy and establishing a socialist political civilization are an important goal for building a well-off society in an all-round way, and that . . . China must go on steadily and surely with political restructuring, extend socialist democracy and improve the socialist legal system."[240] In other words, economic modernization was intimately linked to popular support, a link noticeably problematical in at least one major part of the energy infrastructure.

The PLA is also tied to the nation's hunger for non-domestic energy resources in other ways. The demand for reliable petroleum supplies remains as acute for world powers today as it did a century ago, but with some important differences. Although China has been a net energy importer for a decade, the country retains more than adequate supplies to meet all conceivable PLA missions to defend Chinese vital national security interests.

An interesting comparison may be drawn between present-day China and Great Britain just a century ago, a world power completely dependent on imports for petroleum. This dependence led Britain into imperialist ventures to secure such resources, notably in Mesopotamia (now Iraq) and Central Asia (especially the Baku area, now Azerbaijan). Ironically, these two petroleum-rich regions remain at the apex of international intrigue and potential warfare a century later.

First, the international political situation at the beginning of the 20th century focused on conflict between nation-states in competition for prospective provinces and colonies. Trade was ascendant and linked directly to a nation's military—primarily naval—might. The economic balance is certainly a vital part of international relations in East Asia today, but the globalization phenomena currently on the rise has greatly reduced the emphasis on military, especially naval, power as a necessary ingredient in that process. This is also due to the historically unique position of the United States as the world's dominant military power, with a navy that effectively performs traditional maritime missions, such as SLOC protection, for all Asian nations.[241]

Second, a naval race was a primary feature of the deteriorating relationships among European powers, the United States, and Japan, with even China deploying a small but modern battleship navy. Currently, naval force is important in Asia's maritime environment but is a tool rather than the focus of international competition; nothing like the early 20th-century Anglo-German naval race is taking place. Potential conflicts in East Asia do not focus on territorial colonization, although there are more than 30 territorial disputes in the region, ranging from minor maritime border disputes to the far more serious question of Taiwan's relationship with China.

Beijing has signed border agreements or negotiating regimes with all 14 nations that share its borders, but it still has sovereignty disputes with Japan (over the Daiyutaos [Senkaku Islands to Tokyo]), India, Vietnam, Malaysia, Brunei, the Philippines, and Indonesia over the various land formations and ocean area of the South China Sea. China's attitude toward these sovereignty disputes has for the most part—certainly since early 1996—been moderate; although not conceding any of its claims,

Beijing has sought to establish bilateral and multilateral agreements to contain these disagreements.[242]

Third, national militaries at the dawn of the last century were launching dramatic technological advances that would contribute to a revolution in military affairs (RMA). This would include the development of turbine-driven warships, submarines, aircraft, and armored vehicles, all energy-intensive. An RMA may also be occurring at the dawn of the current century. If so, it is one that will be based on information rather than energy. China's interest in securing energy supplies springs from economic rather than military motivation.

The PLA thus can count on China's indigenous petroleum supplies to fuel its platforms; another resource for the PLAN is nuclear power, already used in six of its submarines.

Other PLA missions remain that affect China's energy needs in the category of national security rather than naval necessity. These missions include, before all others, the requirement to pacify and safeguard civil peace in contentious, potentially resource-rich provinces in the west and northwest. Xinjiang is the most troublesome, but Beijing will likely categorize all non-Han groups (including Tibetans and Muslims residing elsewhere in China) as terrorist, as it has with the ETIM, should members of these groups be perceived as threatening public order. The concern for domestic peace, although in theory primarily dealt with by civilian police and PAP forces, depends in the final analysis on the PLA.

Controlling the ethnic minorities is eased by the fact that they compose, in total, no more than 10 percent of China's vast population, but it is complicated by the fact that many of these groups reside in border areas that historically have been troublesome, such as Xinjiang with the Islamic Central Asian republics, Tibet with India, and Mongolia with Russia. Hence, ethnic irredentist or independence movements are almost certain to involve groups in neighboring nations and trans-border power projection. Beijing will no doubt attempt to enlist third-party, surrogate, or international organizations in any campaign to maintain its claimed territory.

Recent examples include, first, the aforementioned agreement with the United States (and the UN) about the ETIM. The second example is Beijing's efforts to establish a Central Asian security organization as part of the Shanghai Cooperative Organization (SCO). This group was nominally targeted on improving economic relations among its members but more importantly, from Beijing's view, on resolving border disputes, strengthening borders, and establishing international controls over actual and potential cross-border political and religious movements. Since becoming the SCO in 2000 when Uzbekistan joined, and especially since September 11, the group has increasingly focused on security issues, specifically "cooperation against terrorism, separatism and extremism."[243] Cross-border cooperation between national militaries has steadily increased, to include recent military exercises among Chinese and Kyrgyz, and Chinese and Russian, forces.[244] After signing border resolution agreements with all SCO members, Beijing wants further to strengthen the organization, but has met with limited success. The organization announced in November 2002 that an antiterrorism center would be established in Bishkek, Kyrgyzstan, with a secretariat in Beijing.[245]

Furthermore, it must have been very sobering to see the United States, immediately after the events of September 2001, so quickly and apparently effortlessly establish a strong diplomatic and military presence in the very Central Asian region the SCO was designed to control for China. Despite Washington's claims about its antiterrorism focus, there is an undercurrent of suspicion in Beijing that the American presence in Central Asia will not soon end.[246]

Nonetheless, the SCO represents Beijing's most significant effort at multilateralism and delineates the theater most likely to demand PLA missions in the realm of protecting energy resources. Should Uighur separatists expand their terrorist actions, Xinjiang's energy resource infrastructure, including the Tarim Basin fields, might be a likely target.

A presumably more vulnerable target would be the pipeline system Beijing is currently constructing. More than a half-dozen major pipeline projects are built, under construction, or in the planning stage.[247] These fall into two broad categories. The first type is intended to facilitate the distribution of oil and natural gas within China, primarily bringing oil and natural gas from western and northern sources across the country to Shanghai and the rest of economically expanding eastern China. The national plan includes three vertical trunk lines, from Russia, from China's northwest, and the Zhongwu line.[248] The most notable of these trunk lines completed to date is the 775-mile-long line from Lanzhou to Chungqing, built at a reported cost of $500 million.[249] Pipeline security concerns are being addressed by Beijing; the West-East Pipeline's origin in Xinjiang makes it vulnerable to Uighur separatist attention; one CCP official has stated that "The pipeline is going underground to escape both the extreme climate and its potential as a target for terrorist attack."[250]

The second category aims to access foreign energy resources, most immediately those in Russia and Kazakhstan. Other Central and even Southwest Asian sources, including Turkmenistan, may be targeted. As demonstrated by T.E. Lawrence in the Middle East during World War I and currently by the Ejército de Liberación Nacional in Colombia, pipelines are difficult to protect.

Another concern is the possibility of petroleum imports being halted by attacks at sea during the long transit from Southeast Asia, Southwest Asia, the Middle East, or other foreign sources. The SLOCs are most vulnerable not on the high seas, but at transit points through narrow straits, including Hormuz, the 9-Degree Channel, Malacca, Luzon, and Taiwan. The U.S. Navy will protect these SLOCs for the foreseeable future, but a Sino-American crisis (over Taiwan for instance) might drive Beijing to decide that the PLAN had to be capable of defending these SLOCs. The way for China to preclude this eventuality is to resolve Taiwan's status peacefully and to develop continental pipelines as the primary avenue for accessing foreign oil sources. Failing that, Beijing would have to make a major change in national budgeting priorities to build a navy and air force capable of protecting the extended SLOCs that carry much of China's imported oil and natural gas. This is not likely to happen, for several reasons.

First, Beijing retains national priorities that fall under the rubric of "rich country, strong army." That is, developing China's economy and ensuring the welfare of its people will remain the top priority of the national government and the CCP. Second, while the Taiwan issue remains the most sensitive one between Beijing and Washington, the present economic and political situation on the island, U.S. and Chinese interest in keeping the issue within peaceful bounds, and the common interest in the campaign against terrorism all mitigate against the reunification issue deteriorating to the point of hostilities. Hence, Sino-American relations should remain essentially peaceful, if frequently contentious. Third, there is little indication that the Chinese military's strategic paradigm is going to change significantly in the near future. The PLA remains dominated by the army, with the navy only as strong as specific maritime-associated national interests justify.

The Chinese navy currently includes fewer than 20 warships capable of operating in even a limited early 21st-century naval environment. And these ships—2 *Sovremenny*-class, 1 *Luhai*-class, and 2 *Luhu*-class guided-missile destroyers (DDGs), and approximately 12 *Jiangwei*-class guided-missile frigates—are armed with very limited antiair warfare weapons systems. Another 40 or so surface combatants are armed with antisurface ship cruise missiles and, in a non-air threat environment, could perform SLOC defense duties in Chinese littoral waters. The PLAN ability to deploy is further limited by the presence of only three replenishment-at-sea ships in the fleet.

The navy's real strength lies in its numerous, modernizing submarine force. The five nuclear-powered *Han*-class attack submarines are capable of extended deployments but are noisy and difficult to maintain. China's most modern submarines, four *Kilo*-class and three *Song*-class conventionally powered attack boats, are not well suited for long-range deployments (to the Indian Ocean, for example) but are formidable weapons systems within about 1,000 miles of China's coast. Beijing continues to build *Song*s and buy *Kilo*s from Russia; as these boats become operational, the 30 or 40 older *Ming*- and *Romeo*-class boats will be decommissioned.[251]

Since 1983, the PLAN has periodically deployed two- or three-ship task forces on diplomatic missions to Southeast, South, and Southwest Asia and to the Western Hemisphere. In 2002, a *Luhu*-class DDG and an oiler completed a circumnavigation of the globe, a significant accomplishment.[252]

Hence, PLAN is capable of defending littoral SLOCs (those lying no more than 200 nautical miles [nm] from China's coast). Even that capability must be qualified, however, given the proven difficulty of defending surface ships against submarine attack.[253] Defense of China's economic offshore infrastructure is a prominent concern of naval thinking. The South China Sea would become an area of primary PLAN operations should significant energy resources be discovered in waters claimed by Beijing in that sea. PLAN forces have regularly deployed to the Paracel Islands since the early 1970s and to the Spratly Islands since the early 1980s.[254] A Chinese military presence has been established on more than a half-dozen of the islands.

The current maritime strategy is one of offshore defense, meaning that the PLAN will strive to "maintain control over the maritime traffic in the coastal waters of the mainland" and the resources in those waters.[255] Defining this area of capability is not easy, but perhaps the most reliable approach is to look at specific missions and sea lines. This approach yields formidable ocean areas for the PLAN to defend: all of the South China Sea, the western half of the East China Sea, the waters extending from the Chinese coast to at least 100 nm east of Taiwan along a line from the Philippines to Japan, and all of the Yellow Sea.

The next step in this concern is more distant SLOCs, ranging from the Malacca Strait between the South China Sea and the Indian Ocean, and the Hormuz Strait from that ocean into the Persian Gulf. Assuming a continued constructive relationship with the nations of Southeast Asia, China should not have to be concerned with commanding the seas of the narrow Malacca Strait, and the Strait of Hormuz is so far from China that deploying PLAN units to defend it would be impractical (except possibly as part of a multinational force).

A multinational effort is under way to combat the colossal rise in the number of pirate attacks in Southeast Asian waters, attacks that cause an estimated annual loss of $25 billion. China is not playing an active role in the antipiracy center established in Kuala Lumpur by the Association of Southeast Asian Nations (ASEAN), with participation by Japan and India.[256]

As for the vast Indian Ocean distances between the two straits, China faces a wary India with a formidable navy of its own. Beijing's close relationship with Pakistan is marked by significant military assistance to a navy that also is able to count on French submarines and other foreign assistance. Pakistan's force of seven modern, conventionally powered submarines is augmented by eight frigates—none of them new but most armed with guided missiles—and two replenishment-at-sea oilers.

China is also helping Pakistan build a deepwater port at Gwadar, nominally for commercial traffic.[257] But Islamabad has consistently come out second-best in wars with New Delhi, and the advent of the two nations as nuclear powers casts future contests in a different light, especially as India's only way of effectively threatening China is with its nuclear arsenal.

Beijing has begun to establish a military presence in the Indian Ocean and hopes its close relationship with Islamabad will allow it to count on the Pakistani navy in a regional maritime contest. China also has established a strategic economic and military relationship with Burma/Myanmar, the latter by providing advisors and material assistance. The Chinese military and contractor personnel in that country—involved in projects ranging from road-building in the far north to manning listening stations in the Andaman Sea—represent the first Chinese military presence on foreign soil since the Vietnam War.[258]

Several factors motivate Beijing's policy in Burma. First is concern for their common border, historically one rife with drug traffickers and other smugglers, and at one time a refuge for former Nationalist soldiers. Second is the desire to counter Indian influence in the region—important because of its location between the subcontinent and Southeast Asia, an area to which Beijing is devoting increasing political and economic resources.

Third is concern for the Indian Ocean SLOCs on which China depends for so much of its energy imports. At the same time, India is trying to establish a stronger political and naval presence east of Malacca, evidenced in New Delhi's increased attention to ASEAN and the 2001 deployments by the Indian Navy to East Asia, from Singapore to Japan.[259] These events, combined with Indian naval strength in the Indian Ocean, pose a classic problem in maritime strategy for Beijing: its most important source of petroleum imports, the Persian Gulf area, lies at the end of very long SLOCs that are dominated by the navy of a potential enemy.

The Indian Navy continues to modernize and expand. It currently includes an aircraft carrier and at least 17 conventionally powered submarines, with several newer models under construction. The navy surface force centers around 8 DDGs, perhaps 15 older destroyers and frigates, and 3 replenishment-at-sea ships. More importantly, India has funded an ambitious plan to modernize its navy, from new aircraft carriers to submarines.[260]

The naval picture of the Indian Ocean, apart from the usual American presence of one or two aircraft carrier battlegroups, is dominated by an Indian force much stronger than its Pakistani opponent. The PLAN is stronger than either, but it is severely limited by the distances involved.

How will China address the problem of Indian Ocean SLOCs? Beijing apparently has decided not to build a navy capable of patrolling these long SLOCs to the Middle East. Instead, Beijing seems to be concentrating on forming supportive relationships with the nations bordering those routes, from the Philippines to Saudi Arabia. Beijing's concern for the security of its overseas energy supplies need not dominate its national security policy process. Rather, the most important aspects of energy security are economic and political, not military.

Given China's aforementioned significant draw on Middle Eastern-Southwest Asian oil, a prolonged war in that region might well seriously disrupt the outflow of petroleum products. To forestall or ameliorate that eventuality, Beijing is engaging in diplomatic activity both to signal its interest in the welfare of the Arab states and to offer mediation services in the Israeli-Palestinian conflict.[261] This activity backs up and possibly extends Beijing's activities with petroleum companies in the region, including investments or extraction activities in Iran, Iraq, Kuwait, Saudi Arabia, Egypt, Sudan, and Somalia.

The maritime oil flow to China would be subject to unrest in Southeast Asia but almost certainly would not be seriously affected. For one thing, the political situation in that region is so fractured as to make effective international action against freedom of navigation extremely unlikely. For another, even if the Malacca Strait–South China Sea route was interrupted, oil could be shipped via alternate routes at an acceptable increase in cost.[262]

Chapter Ten

Conclusion

A lmost all observers forecast that China's economy will continue to grow, a process that will require annual energy increases of 4 to 5 percent through 2020 (compared with growth of about 1 percent in the industrialized countries).[263] China currently both consumes and produces about 10 percent of the world's energy.[264]

China will increasingly be a net energy importer, although the nation will remain a net exporter of coal through the forecast period. Coal will remain China's primary fuel for generating electricity; hydroelectricity, natural gas, and nuclear energy will become more important, while the oil sector will decline, losing at least half its share.

The natural gas share of total energy production is expected to double by 2020 as China begins to take greater advantage of its large domestic reserves. Beijing has vowed to cut pollution by "increasing the proportion of gas and electricity in energy consumption [by] 75 percent in 2005, and 83 percent in 2010."[265] The nuclear share of power generation may quadruple but will still form just 1.6 percent of production, while the hydroelectric share is expected to increase by a third, as the petroleum share falls.

China's total energy production, 29.4 quads in 1990, had increased to 34.9 quads by 2000. By comparison, the United States produced approximately 70.9 quads of energy in 1990 and 99 quads in 2000. By 2015, China's overall energy production is expected to exceed 70 quads, of which approximately three-fourths will come from coal.

On the other side of the ledger, China consumed 27 quads of energy in 1990, representing about 9 percent of world energy consumption. Of this amount, coal accounted for approximately 73 percent and petroleum another 20 percent. U.S. consumption in 1990 was more than 84 quads; by 2000, that number had risen to almost 99 quads.

The industrial sector typically consumes approximately 75 percent of China's energy annually. Continuing increases in industrial energy efficiency are expected from such measures as installation of more efficient boilers, but industry is also likely to become more electricity-intensive as it phases out direct fuel burning. The transportation sector, by contrast, accounts for only about 7 percent of China's energy consumption. This amount will likely increase rapidly, however, as private automobile ownership increases.

China's electricity demand more than doubled between 1986 and 1995 and is expected to triple between 1995 and 2020. Production of electricity is forecast to match this figure.[266] Since only 80 percent of Chinese citizens are connected to an electrical grid (quite low compared to most industrialized nations), the residential/commercial sector should experience the most rapid growth in electricity demand, driven largely by enormous increases in appliance ownership and continued electrification of rural areas.

China will continue exploring for additional domestic oil and gas resources ashore, with the northwestern provinces the most likely source—despite relatively disappointing results to date.[267] The goal to reduce the overwhelming reliance on coal for power generation has spurred a campaign to increase dramatically the use of natural gas. This goal also aims to lower reliance on imported oil, reduce the damaging environmental effects of burning coal, and increase the efficiency of the energy sector.

Tied to this aim is the fact that domestic oil production may not increase significantly in the immediate future. The Daqing oilfield, long China's most productive, is suffering from decreasing supplies and inefficient reorganization; oil prices are low and no great new petroleum fields have emerged in China; the transportation infrastructure for energy products remains constrained; and the country is in the midst of economic reorganization that roils the energy sector picture. The Minister of Land and Natural Resources stated in April 2002 that the country's "oil and gas reserves are in great shortage" and that "the oil fields in eastern China are unable to increase yield any more."[268] Furthermore, the Ministry of Land and Resources reported that "for the past ten years, China's crude oil consumption has kept growing at an average rate of 5.77 percent, while the growth of domestic oil supplies was only 1.67 percent."[269]

Energy is an increasingly significant factor in Beijing's national security calculus and hence in the conduct of foreign relations. Future maritime exploration for energy resources is a significant aspect of this fact. China has for many years carried out an extensive program of maritime exploration and surveying. In many cases these surveys serve both national security and economic objectives; mapping the ocean bottom, for instance, yields data useful for future submarine and antisubmarine warfare applications, as well as for locating likely areas of petroleum reserves. To improve this program, Beijing has begun an effort to establish a system of navigation and surveillance satellites.[270]

These operations have caused diplomatic protests and influenced domestic political discourse in several Asian nations. Surveying missions in waters claimed by Japan have prompted frequent complaints by Tokyo. Similar operations in the South Pacific have evoked protests from Vanuatu.[271] In other words, although a signatory of the 1982 UNCLOS and quick to protest violations of that document's parameters, Beijing does so based on its almost unique interpretation of maritime law and does not hesitate to exploit that interpretation for its own purposes.

The ocean-bed resources from which China currently draws petroleum products are located in territorial waters or those well within the nation's exclusive economic zone, for the most part within 100 nm of the mainland. The only likely change to this situation would occur if significant oil or gas reserves are discovered in the middle or southern part of the South China Sea or well out in the East China Sea, which would extend Beijing's proven energy interests out as far as 750 nm from China.

The most promising additional maritime energy deposits are likely to be either natural gas or methane hydrates. The latter is a gas found in solid form beneath the ocean bottom. Economic recovery of significant quantities remains beyond current technology, although since the early 1980s the United States, Russia, Japan, India and several other countries have experimented with its recovery.

As an example of its potential, possible methane hydrate resources in U.S. waters are estimated (at 50 percent probability) at 9,000 tcm, which is almost 300 times larger than the estimated total remaining conventional natural gas resources in the United States.[272] Should this resource be found in Chinese waters, the implications for a net energy importer like Beijing would be significant. Some Chinese engineers have estimated that huge amounts of methane hydrate reserves lie on the Tibetan Plateau, as well as in the Bo, East China, and South China Seas, but the theoretical rather than proven nature of these reserves is described by Dr. Jin Xianglong, a researcher with China's State Laboratory of Ocean Floor Science under the State Bureau of Oceanography.[273]

The efficient transfer of natural gas to market (on mainland China) requires liquefaction, given the current limits for piping natural gas. That in turn means that Beijing would require the cooperation of the nations closest to the gas deposits to build liquefaction plants. Hence, even the discovery of large, economically recoverable energy resources in the mid- to southern South China Sea is not likely to evoke military action by Beijing.

There is no doubt about the government's deliberate, well-funded global effort to locate and secure energy supplies. This search is spurred by the lack of success locating additional domestic petroleum resources and the probability that offshore sources will never satisfy more than approximately 10 percent of China's petroleum requirements. Hence, Beijing will remain forced to seek international energy supplies.

During 2002, CNPC engaged in exploration and production in Russia, Libya, Syria, Algeria, Iran, Iraq, Oman, Tunisia, Kazakhstan, Venezuela, Burma, "and other Southeast Asian countries."[274] China is also actively seeking oil imports from other nations in Latin America, sub-Saharan Africa, Central Asia, and the Middle East. As discussed above, Beijing has petroleum products purchase agreements with several countries in these regions.[275] Another aspect of this policy is China's reported attempt to purchase outright one of Russia's state oil firms.[276]

China faces two serious difficulties in Latin America, one immediate and one more general. The first is the political uncertainty in Venezuela, the Latin American nation with which Beijing has signed its most significant energy acquisition agreement. The turmoil of President Hugo Chavez Frias' regime highlighted severe economic, political, and social problems common to all too many Latin American nations, to which are offered few solutions other than to blame the United States. His agreements with Beijing resulted more from a desire to tweak Washington than to form a lasting relationship with China.

This leads directly to the more general problem China faces in acquiring Latin American energy resources: the United States remains the dominant economic and political presence in the hemisphere, with European interests in second place. It is difficult to imagine circumstances that would permit China to acquire significant energy resources from Latin America.

Beijing faces a similar problem in sub-Saharan Africa. The oil-rich nations of this region are already heavily tied to Western oil countries, including the United States, Great Britain, the Netherlands, and Brazil. This does not mean Beijing could not make inroads on the continent, but it does mean that such efforts may incur a significant cost in relations with the nations with whom China would be competing. Mozambique is a case in point: Chinese efforts to take part of this state's oil production by supplanting the Brazilian companies already present would affect Chinese efforts to expand its economic presence in South America.

A great many articles have appeared in the open press during the past decade about Russia's Far East serving as the source of large energy supplies for China. The Russian company Yukos has been supplying oil to China by rail since 2001 and recently signed an agreement with CNPC to increase these shipments through 2005.[277]

After at least 2 years of discussion, in July 2001 President Jiang Zemin signed an agreement in Moscow for a feasibility study of a 400,000-bbl-per-day pipeline from Angarsk in eastern Siberia to Daqing, in northeastern China.[278] This project remained in the planning stage as of June 2003, with fall 2003 the most optimistic start date.[279] Available details describe a 1,400-mile-long pipeline capable of providing 147 million bbl of crude oil a year to China when completed in 2005, and 221 million bbl by 2010.[280] This proposed pipeline, which would cost approximately $2 billion, would have the capacity to convey 30 million tons of petroleum—a very significant proportion of China's total demand—from Russia to China annually by 2010. The lead companies for the project would be CNPC and PetroChina for China, and Yukos and Transneft for Russia.[281] The two governments have also discussed the sale of Russian surplus electricity to China.[282]

Despite all the talk—and the almost certain presence—of large natural gas and oil deposits in Siberia, very little production from this area has shipped to China. One reason for this is that Russia's energy exporting infrastructure is oriented toward Western Europe. Redirecting that infrastructure to supply China would be a large undertaking based on political as well as economic evaluations.

This situation evokes the decades of desultory reports of Japanese plans to provide the investment necessary to recover the mineral wealth of Russia's Far East. The wealth is there, but in China's case, the necessary technological and financial resources, not to mention the political will, so far have been lacking. Furthermore, Moscow's desire for the financial return from selling these resources to Beijing is shadowed by historic Russian doubts about Chinese reliability.

As of March 2003, the discussions between Moscow and Beijing remain essentially stalled, with Russia negotiating with Japan about an alternative: instead of building a pipeline from Angarsk in Siberia to Daqing in northeastern China, Moscow would run the pipeline from Angarsk to Nakhodka, a seaport that would facilitate the shipment of Russian oil both to Japan and to other Pacific nations. Moscow may originally have aired this idea to pressure Beijing into an agreement about exploiting Siberian reserves, which are plentiful but expensive to recover, but Tokyo's proposal to share the $5 billion cost of the pipeline strongly appeals to Moscow. The most likely result of the two pipeline proposals, which have engaged Beijing's testy attention, is that a pipeline from eastern Siberia will be built with branches both to China and to Nakhodka. Moscow has yet to announce its decision.[283]

Beijing apparently is making an extensive effort to include the energy sector in any strategic partnership with Moscow, and a similar effort is being made with Russia's former republics in Central Asia, with the SCO as vehicle. Extensive programs have been launched in Kazakhstan and Kyrgyzstan, with whom Beijing has signed agreements and contracts and from whom it has purchased a small amount of oil. China's energy policy in Central Asia is shadowed by competition from Russian and Western oil companies seeking to recover the area's potentially rich petroleum deposits, and increasingly by the extensive post-September 11 American presence on China's western border.

China's rulers historically have viewed Central Asia as the source of threats to national security. Beijing no doubt looks favorably at the suppression of the transnational Islamic movements in Central Asia that have affected its rule in Xinjiang, but it also must have a good deal of concern about both the renewed Russian interest in its former republics and the U.S. indications of maintaining a strong presence in the region for the foreseeable future.

The proposals for various trans-Asian pipelines would require international cooperation and, probably, sharing of energy resources on a scale demanding Beijing's agreement to a dramatically new degree of economic cooperation and opening. These proposals include an ASEAN Gas Center; a pipeline that would serve Russia, China, Japan, and Korea; and a visionary complex of pipelines that would stretch from Kazakhstan to Shanghai and from Australia to Russia's Far East.[284]

This project assumes a 58.8 percent growth in Southeast Asia's energy needs by 2010 and would require an investment of an estimated $180 billion over the next 10 years.[285] As a participant in ASEAN Plus Three discussions, China may be part of cooperative ASEAN efforts and has held talks on energy security in Asia with Japan and South Korea.[286] The Asian energy equation is clouded by estimates that two of the region's largest producers, Indonesia and Malaysia, will exhaust exportable resources by about 2012 and 2016, respectively.[287] Over 60 nations met in September 2002 in Osaka to address uncertainties in Asia's energy supply. China agreed with Japan, South Korea, and the ASEAN nations to establish an information-sharing network as part of preparing a set of emergency response measures.[288]

China's leadership recognizes the importance of energy resources in their nation's economic, social, and political future. In October 2001, the SETC called for the "rectification and standardization of the [oil] market." Standardization was to be enforced at all levels of government, from national to municipal, and at all levels of the economy, from wholesale importation and storage to individual gas stations.[289]

Beijing is focusing on the nation's energy requirements, supplies, infrastructure, environmental concerns, and international ties, but it is still struggling with the question of how much state control to exert over the nominally private energy sector. Numerous government statements, organizational moves, and actions by state-sponsored "private" energy conglomerates all dance around this issue: how freely will "socialism with Chinese characteristics" allow the energy system to function?

This issue can be difficult to resolve, as evidenced in the widespread labor unrest that has accompanied recent privatization efforts. The maximization of foreign investment of capital, technology, and developmental skills is also recognized as a requirement, as is expanded access and even outright acquisition of a variety of energy resources, literally on a global basis. One enlightening report describes the Liaohe River Petroleum Exploration Bureau, a state-owned oil company in troubled Liaoning Province. In the process of privatizing, the company significantly reduced employment, cutting well-drilling and production teams "in accordance with market demand." It is also "targeting the international market," with projects in North Africa, South America, and Russia. This report claims success in economic terms but does not explore the resulting unemployment problems.[290]

Another aspect of China's energy sector that probably troubles Beijing is the reliance on foreign companies for exploration, exploitation, and marketing. Foreign investment in China's energy sector is concentrated in the hands of American and Western European companies, which raises interesting possibilities for non-national influences on Beijing's future national security decisions.

China's ability to control its rapidly changing energy situation is marked not only by dependence on foreign technology and investment, but also by the mining industry's terrible safety record, the societal unrest resulting from corruption and from the privatization of SOEs, the lack of an adequate transportation and pipeline infrastructure, the weakness of the domestic governmental system to coordinate among national and provincial governments, and the ideological manacles of "socialism with Chinese characteristics."

As evidenced in the past decade's new willingness to engage in multilateral forums, conduct land and maritime border negotiations, and modernize the military, China's long-term campaign to secure energy will continue to affect the pursuit of national security and foreign relations. If China's energy requirements continue to grow as forecast and significant new domestic sources are not found, then reliance on foreign energy resources will increasingly be a factor in Beijing's policy options.

Dependence on foreign energy supplies may result in a studied departure from the five "peaceful coexistence" principles so long adumbrated by Beijing, if foreign energy resources face political or economic threats. As long as the CCP governs China and believes continued economic growth essential for regime survival, then Beijing will use all instruments of statecraft, military as well as economic and political, to ensure adequate energy supplies are available.

This does not mean that severe domestic unrest or international conflict is inevitable or even likely. Rather, it means that as China becomes more integrated into the international economy through the vehicle of the WTO, the energy sector will play a more significant role in both its domestic and foreign policy priorities. Hence, other nations dealing with China will find themselves focusing increasingly on this part of their relationship, while the Chinese government will have to react to greater pressure to liberalize its political rule over a liberalized economy.

Endnotes

[1] Alice Tisdale Hobart, *Oil for the Lamps of China* (New York: Bobbs-Merrill, 1933).

[2] "China," Department of Energy-Energy Information Agency (DOE-EIA) Country Analysis Brief, April 2000, accessed at <www.eia.doe.gov/emeu/cabs/china.htm>, 1.

[3] Cited in Felix K. Chang, "Chinese Energy and Asian Security," *Orbis* 45, no. 2 (Spring 2001), 213.

[4] Statement to author by senior U.S. Department of Energy official, December 2001.

[5] "China: Environmental Issues," DOE-EIA, October 1999, 2.

[6] These numbers are probably approximately correct, but it is easy to find different estimates for the energy sector. For instance, 17.4 percent is given as the percentage of China's total power output for 2001 provided for hydropower in "Profile of China's Electric Power Industry," *Asia Pulse* (January 6, 2003), in *Alexander's Gas and Oil Connections* (*Alexander's*) 8, no. 2 (January 24, 2003).

[7] "China Drafts Energy Development Strategy for 21ˢᵗ Century," *Xinhua* (Beijing edition, unless otherwise indicated), May 23, 2000, in Foreign Broadcast Information Service (FBIS)-CPP20000523000134. As will be discussed below, the other energy sources range from biomass to tidal power and wind power.

[8] "China to Rely Increasingly on International Markets for Its Oil Supply," *Xinhua*, in *Alexander's* 6, no. 15 (August 14, 2001), accessed at <www.gasandoil.com/goc>.

[9] "Beijing Has No Plans to Depart From Foreign Investors Schedule," Reuters, in *Alexander's* 6, no. 15 (August 14, 2001).

[10] Table 2, "East Asia: The Energy Situation," DOE-EIA, August 1999, 5.

[11] 36 million tons in 1999 and 70 million tons in 2000; reported in Zhu Yuan, "Sino-Russian Oil Deal," *China Daily*, September 11, 2001, in FBIS-CPP20010911000131; the 2002 figures are from the Chinese Customs Service, reported in *Takungpao News*, February 11, 2003, in *Virtual Information Center*, "Asia-Pacific Daily News Summary."

[12] "China Increasingly Reliant on Imported Fuel Supplies," Associated Press, quoted in *Alexander's* 5, no. 14 (August 7, 2000); "China," DOE-EIA, 1

[13] "PRC to End State Monopoly, Urges Competition in Utility," *Xinhua*, November 21, 2000, in FBIS-CPP20001121000079.

[14] Other works that address this topic include Selig S. Harrison, *China, Asia, and Oil: Conflict Ahead?* (New York: Columbia University Press, 1977); Yingzhong Lu, *Fueling One Billion: An Insider's Story of Chinese Energy Policy Development* (Washington, DC: Washington Institute Press, 1993); Mark J. Valencia, *Southeast Asian Seas: Oil Under Troubled Waters: Hydrocarbon Potential, Jurisdictional Issues, and International Relations* (New York: Oxford University Press, 1992), and *Conflict Over Natural Resources in Southeast Asia and the Pacific* (New York: Oxford University Press, 1996); Xiannuan Liu, *China's Energy Strategy: Economic Structure, Technological Choices, and Energy Consumption* (Westport, CT: Praeger, 1996); Mark J. Valencia, Jon M. Van Dyke, and Lowell A. Ludwig, *Sharing the Resources of the South China Sea* (Honolulu: University of Hawaii Press, 1999); Erica Strecker Downs, *China's Quest for Energy Security* (Santa Monica, CA: RAND, 2000); Robert A. Manning, *The Asian Energy Factor: Myths and Dilemmas of Energy, Security and the Pacific Future* (New York: Palgrave, 2000); Felix K. Chang, "Chinese Energy and Asian Security," *Orbis* 45, no. 2 (Spring 2001),

211–240; Amy Myers Jaffe and Steven W. Lewis, "Beijing's Oil Diplomacy," *Survival* 44, no. 1 (Spring 2002), 115–134.

[15] "Organization," in "China: An Energy Sector Overview," DOE-EIA, December 1999.

[16] Zheng Bijian, "Fundamental Trend of Socialism with Chinese Characteristics in the New Century," *Renmin Ribao* (Beijing edition unless otherwise indicated), November 21, 2002, in FBIS-CPP20021121000023.

[17] Manning, 107–109.

[18] Jing Ji, "Sinopec's Move Stirs Controversy," *China Daily*, November 5, 2002, in FBIS-CPP20021105000045.

[19] "Country Analysis: China," DOE-EIA (June 2002), in *Alexander's* 7, no. 13 (June 27, 2002).

[20] "Sudanese Energy Minister Meets Chairman of Chinese National Oil Company," Suna News Agency, August 15, 2002, in FBIS-AFP20020816000160, indicates that current cooperation is minimal.

[21] "China Launches Energy Development Project," *Xinhua*, August 25, 2001, in FBIS-CPP20010825000066.

[22] "China Maps Out Plan for Overall Development of Industrial Sector," in *Alexander's* 6, no. 16 (August 28, 2001). Also see "New Energy Sources Highlighted in China's Strategy for Next Five Years," *Xinhua*, August 10, 2001, in FBIS-CPP20010810000131, for the report that "more foreign capital will be used and foreign cooperation will be expanded" in the effort to draw on "new [and renewable] energy resources." The four goals are discussed in *Asia Pulse* (August 2001) and "China Releases Development Plan for Power Industry," in *Alexander's* 6, no. 15 (August 14, 2001).

[23] "China Will Boldly Go Into International Energy Exploration," *China Electrical News*, in *Alexander's* 6, no. 15 (August 14, 2001). Also see "Hydroelectricity and Other Renewable Sources," DOE-EIA Report 0484 (2000), accessed at <www.ei.doe.gov/oiaf/archive/ieo00/hydro.html>, for the World Bank report.

[24] Frank Umbach, "China's Energy Policy," *Internationale Politik* (Transatlantic Edition) (Summer 2001), 88.

[25] Wang Xianzheng, quoted in Jin Tian, "Energy Structure Calls for Fine-Tuning," *China Daily* (Business Weekly Supplement), June 18, 2002, in FBIS-CPP2002061800063. Wang has since become vice governor of Shanxi Province—China's largest coal-producing province.

[26] Coal is measured in "short tons," unless otherwise noted. One short ton equals 2,000 pounds (or 908 kg or 0.89 long tons or 0.91 metric tons).

[27] "Coal in the Energy Supply of China: Executive Summary," November 1999, accessed at <www.iea.org/pubs/studies/files/coachi/coachi.htm>, 1–3.

[28] Table A–6, "World Coal Consumption, 1990–2020," in "International Energy Outlook 1999," DOE-EIA, accessed at <www.eia.doe.gov/oiaf/ieo99/tbla1-8.htm>, 9. No other nation exceeds 400 mst in 2000; U.S. coal consumption in 2020 is estimated at 1,275 mst.

[29] Ibid., 3. Increased coal exports are discussed in "China's Coal Export Up 45.6 Percent in 2000," *Xinhua*, July 6, 2000, in FBIS-CPP20000706000010; also see Gong Zhengzheng, "Oil Prices to Fire Up Domestic Coal Exports," *China Daily* (Business Weekly Supplement), April 30, 2002, in FBIS-CPP200204030000073.

[30] The foregoing discussion relies heavily on DOE-EIA, "Country Analysis Brief of China." "China, U.S. to Jointly Develop Coalbed Methane," *Xinhua*, March 3, 2003, in FBIS-CPP20030303000173.

[31] Reported in "China and Japan Reach Oil and Coal Framework Agreement," Reuters, December 14, 2000, in *Alexander's* 6, no. 1 (January 11, 2001).

[32] "Accidents in Hunan, Guangxi Claim Lives of 41 Miners," *China Daily* (Hong Kong), January 9, 2001, in FBIS-CPP20010109000029.

[33] The smaller number is in Jiang Zhuqing, "Thousands of PRC Illegal Mines Shut Down; Billions Invested to Improve Safety," *China Daily*, April 10, 2002, in FBIS-CPP20020410000019. In "China to Shut Down More Small Coal Mines, Decrease Accidents by 10 Percent," *Xinhua*, April 10, 2002, in FBIS-CPP20020410000001, a "government official" announced a goal for 2002 of reducing the total number of illegal mines to 15,000. The larger number is cited in Jin Tian, 2.

[34] The government figure is cited in Jasper Becker, "PRC Issues New Health, Safety Regulations for Mining Industry," *South China Morning Post* (Hong Kong), December 19, 2000, in FBIS-CPP20001219000063. "Thirty-One Trapped After Coalmine Gas Explosion in Northeast China," Agence France-Presse (henceforth AFP), November 6, 2000, in FBIS-CPP20001106000111, cited the 10,000 figure. "Accidents in Hunan, Guangxi Claim Lives of 41 Miners" reports, "Coal mine accidents

have claimed more than 5,300 lives in China in the past year." Also see "Poor Safety Kills 162 Coal Miners in Southwest China," which reports 162 miners dying in a September 27, 2000, mine explosion in southwestern Guizhou Province; and "China Mine Manager Arrested After Explosion Toll Rises to 41," AFP, December 10, 2000, in FBIS-CPP20001210000018, which reports that "also detained were five mine workers accused of beating up journalists and miners . . . in an attempt to cover up the explosion." Also see "East Asia: The Energy Situation," DOE-EIA (August 1999), accessed at <www.eia.doe.gov/emeu/cabs/eastasia.htm>, 3.

³⁵ "Local Authorities Sue Six for Fatal Mine Explosion," *Xinhua*, December 27, 2001, in FBIS-CPP20011227000060, reported that an additional seven persons were being prosecuted by the government, including the "deputy secretary of the [local] Committee of the CCP"; "Nine Missing, Presumed Dead in Coal Mine Explosion in Central PRC," AFP, September 19 2001, in FBIS-CPP20010929000297; "Gas Blasts Kill 37 in Two Coal Mines," *South China Morning Post*, November 17, 2001; accessed at <http://china.scmp.com/ZZZ7JGD81UC.html>; "Police Arrest Four for Mine Disaster," *China Daily*, November 19, 2001, in FBIS-CPP20011119000015; "Small Coal Mines in Shanxi Ordered to Stop Production for Safety," *Xinhua*, November 19, 2001, in FBIS-CPP20011119000096.

³⁶ "Individuals Responsible for 15 November Coal Mine Explosion Receive Prison Sentences," *Xinhua*, December 7, 2001, in FBIS-CPP20011207000068; "Coal Mine Explosion Kills 11 in Central China," *Xinhua*, December 14, 2001, in FBIS-CPP20011214000119; "Fifteen Miners Missing in China Mine Flood," AFP, December 25, 2001, in FBIS-CPP20011225000064; "Coal Mine Blast Kills Eight in Northeast China," *Xinhua*, December 29, 2001, in FBIS-CPP200112290000119; "Gas Explosion in Jiangxi Coal Mine Kills 20, Injures 24," *Xinhua*, December 31, 2001, in FBIS-CPP20011231000039.

³⁷ "Zhu Rongji Signs Order No. 296 of the PRC State Council," *Xinhua*, November 14, 2000, in FBIS-CPP20001114000019.

³⁸ Chapter 4 of the new regulations, in "Apparent Full Text of Regulations on Supervision of Coal Mine Safety," *Xinhua*, November 14, 2000, in FBIS-CPP20001114000032; "PRC: Zhu Rongji Signs Order on Coal Mine Safety Rules," *Xinhua*, November 4, 2000, in FBIS-CPP20001114000019, reported that the premier personally signed the new regulations.

³⁹ "Wu Bangguo Visits 'Disadvantaged' Coal Miners in Shanxi," *Xinhua*, January 14, 2003; "Vice Premier Wen Jaibao Spends Holiday with Coal Miners, Urges Improved Safety," *Xinhua*, February 1, 2003. Wu has since assumed the position of head of the National People's Congress, while Wen has become prime minister, both among the top five leaders in China.

⁴⁰ Jiang Zhuqing reported 2 billion yuan "spent on 103 safety improvement projects," but all in state-owned mines. In the first 3 months of 2002, 27 miners died in Hebei coalmine explosion, 23 were lost in Sichuan, 22 were killed in Henan, at least 24 died in Heilongjiang, and an unknown number died in coal mining accidents in Hubei and Liaoning provinces; see Chu Xuejun, "Major Gas Explosion," *Xinhua* (Hong Kong), April 25, 2002, in FBIS-CPP20020425000083; Yang Haibin and Lin Su, "Heilongjiang Coal Mine Explosion Kills 24," *Xinhua* (Hong Kong), April 10, 2002, in FBIS-CPP20020410000109; also, Bill Savadove, "22 Die in China Coal Mine Blast," *South China Morning Post*, April 2, 2002, in FBIS-CPP20020402000049; "Coal Mine Explosions in Northern China Kill 27," *Xinhua*, January 27, 2002, in FBIS-CPP20020127000069; "Seven Chinese Miners Die of Gas Carbon Monoxide Poisoning," AFP, February 24, 2001, in FBIS-CPP20020224000021; "Three Dead, 19 Missing in Coal Mine Fire in Liaoning," AFP, March 4, 2002, in FBIS-CPP20020304000157; Da Yong, "Coal Mine Disaster Kills 1, Traps at Least 5 Victims in Liaoning," *China Daily*, March 19, 2002, in FBIS-CPP20020319000035.

⁴¹ These accidents are reported in "Death Toll Rises to Eight in Jilin Coal Mine Accident," *Xinhua*, January 11, 2003; "Rescuers Race to Reach 32 Miners," AFP, January 12, 2003; "Gas Leak Kills 18 Henan Province Miners," *Xinhua*, January 24, 2003; "Heilongjiang Coal Mine Blast Traps Six," *Xinhua*, February 25, 2003; "Coal Mine Explosions in Guizhou, Heilongjiang Kill 35," *China Daily*, February 26, 2003; Shih Hsueh-ya, "[Vice Premier] Wu Bangguo Makes Written Comments on Gas Explosion at Guizhou Coal Mine," *Wen Wei Po* (Hong Kong), February 28, 2003.

⁴² "China Reports Rises in Coal Production, Marketing," *Xinhua*, December 29, 2001, in FBIS-CPP20011229000086; "China's Coal Exports Said Profitable for 'First Time in Years'," *Xinhua*, December 24, 2001, in FBIS-CPP20011224000006.

⁴³ "China to Import More Coal," *Xinhua*, February 13, 2003. An American engineer involved in building and operating large, coal-powered generating plants in Fujian Province told the author in

February 2000 that coal could be more cheaply and reliably imported from Australia than from Manchuria.

[44] "First Section of 'Major Coal Transport Railway' Opens in Inner Mongolia," *Xinhua*, December 16, 2000, in FBIS-CPP20001216000078.

[45] "China, U.S. Jointly Develop Coalbed Methane," *Xinhua*, March 3, 2003.

[46] "Underground Vaporization of Coal Passes PRC Ministerial Test," *Xinhua*, November 15, 2000, in FBIS-CPP20001115000016.

[47] See "U.S. Firm Joins Methane Project," *China Daily*, March 4, 2003, "PRC, U.S. Companies to Jointly Explore Coalbed Methane," *Xinhua*, January 8, 2003, and "China Signs Coal-bed Methane Contracts With U.S. Texaco Petroleum Company," *Xinhua*, November 8, 2000, in FBIS-CPP20001108000114, 1. Also see Xu Dashan, "PRC, Canada Sign Memorandum to Develop China's Coalbed Methane Technology Project," *China Daily*, March 29, 2002, in FBIS-CPP20020329000032.

[48] "China, U.S. to Jointly Tap Coal Bed Methane Resources in Yunnan," *Xinhua*, December 3, 2002, in FBIS-CPP20021203000181.

[49] This intriguing program is being led by Lawrence Livermore Laboratory; see John Cooper, "Turning Carbon Directly into Energy," *Science and Technology Review*, June 2001, accessed at <www.llnl.gov/str/June01/Cooper.html>.

[50] Petroleum figures will be given in barrels, unless otherwise noted. One barrel of petroleum/petroleum products equals approximately 42 U.S. gallons or 0.136 metric tons or 0.15 short tons.

[51] These estimates are from "China Expects to Import 50 Million Tons of Oil in 2000," *Xinhua*, November 21, 2000, in FBIS-CPP20001121000078, and "China," EIA, November 1999, accessed at <www.iea.org/about/nmcchina.htm>, 1. They are repeated in the more timely estimate by the International Energy Agency, cited in "China's Oil Demand Expected to Rise," *Dow Jones*, September 11, 2002, in *Alexander's* 7, no. 19 (October 1, 2002).

[52] Various energy sources, in this case petroleum and coal, are difficult to compare in gross terms; more significant is a percentage comparison, which in China reveals a continuing dominance of coal as the most important energy fuel. See, for instance, Gong Zhengzheng, "PRC Analyst Views State Law Proposed to Strengthen Oil Security," *China Daily*, October 30, 2001, in FBIS-CPP20011031000022, for the statement by State Economic and Trade Commission official that "we must quickly form regulations" to secure Sino-foreign cooperation in oil exploration, pipeline protection, overseas oil prospecting, "the State oil stockpile."

[53] "China Busy to Win Security of Supply for a Growing Economy," Reuters, April 27, 2000, in *Alexander's* 5, no. 7 (April 27, 2000).

[54] This joint venture is nominally headquartered in the British Virgin Islands to avoid Taipei's restrictions on cross-strait business dealings. CPC announcement in Luis Huang, "CPC, CNOOC to Sign Agreement on Oil Exploration Soon," Central News Agency (Taipei), in FBIS-CPP20020123000197; also see "CNOOC Finalizes Plans to Explore for Oil and Gas in Taiwan Strait," Reuters, August 28, 2000, in *Alexander's* 6, no. 18 (September 25, 2001), which noted that cross-strait companies jointly conducted geophysical surveys from 1996 to 2000. David Hsu, "Taiwan, Mainland Chinese Oil Companies May Jointly Explore in Strait," Central News Agency (Taipei), in FBIS-CPP20011010000182, dates this agreement to 1998.

[55] "Cross-Strait Oil Deal Delayed," *Taipei Times*, February 7, 2002, in FBIS-CPP20020207000144. A more optimistic report notes that "a final joint-venture agreement" will be signed in May (Xie Ye and Gong Zhengzheng, "JV to Tap Resources Across Straits," *China Daily*, April 19, 2002, in FBIS-CPP20020419000013), while "Taiwan and China Sign Landmark Oil Exploration Pact," Reuters, May 16, 2002, in *Alexander's* 7, no. 12 (June 13, 2002), reports signing of an agreement. "China and Taiwan Undertake Joint Oil Search," *Neftegaz.RU*, January 29, 2003, in *Alexander's* 8, no. 4 (February 20, 2003), reports that the "the two countries, Beijing and Taipei," have approved the dual venture, but the verbiage of the announcement casts doubt on its accuracy.

[56] "CNOOC Signs First Deepwater Exploration Agreement with Husky Oil," CNOOC Ltd., December 6, 2002, in *Alexander's* 8, no. 1 (January 10, 2003). This agreement reportedly requires Husky to pay 100 percent of the exploration costs, while CNOOC would garner 51 percent of any discoveries.

[57] "China Petroleum Completes Historic Kuwaiti Oil Project," in *ChinaOnline*, accessed at <www.chinaonline.com/industry/energy/NewsArchive/Secure/2001/April/C01042305.asp>. "Shenzhen Huxian to Invest $600 Million in Gas Sector," accessed at <www.bangladeshweb.com/news/mar/18/e18032003.htm#A1>, reports the deal with Bangladesh.

⁵⁸ "China's Largest Oil Company to Further Boost Overseas Presence," *Xinhua*, January 15, 2003, in FBIS-CPP20030115000011.

⁵⁹ Umbach, 88. Also see "Turkmen Oil Company Cooperates with China," *Turkmen Press*, November 21, 2002, in FBIS-CEP20021122000186; "China to Invest in Venezuelan Industries," Reuters, in *Alexander's* 6, no. 24 (December 27, 2001); "CNPC Acquires Stake in Myanmar Oil and Gas Blocks," *Platts*, in *Alexander's* 6, no. 24 (December 19, 2001); Xie Ye, "CNOOC Finalizes Acquisition of Indonesian Oil Assets from Spanish Firm," *China Daily*, January 19, 2002, in FBIS-CPP20020119000006; Mai Tian, "Canadian Oil Firm Confirms Discussions, 'Potential Transactions' With PetroChina," *China Daily*, February 21, 2002, in FBIS-CPP20020221000042; "Chinese Oil Experts Begin Oil Exploration Work," Radio HornAfrik, February 14, 2002, in FBIS-AFP20020214000044.

⁶⁰ "China to be Largest Buyer of LNG from Tangguh, Papua," *Timika Pos*, February 8, 2002, in FBIS-SEP20020211000124; partial (12.5 percent) ownership of the field is reported in "CNOOC to Take Stake in Tangguh Gas Fields From BP," *Platts*, September 26, 2002, in *Alexander's* 7, no. 20 (October 15, 2002); also see "CNOOC Acquires LNG Reserve Stocks from BP," *SinoCast*, February 14, 2003, in *Alexander's* 8, no. 5 (March 6, 2003). BP management of the Fujian terminal operation is reported in "BP and Partners to Supply LNG to China's Fujian LNG Terminal," *Power Engineering International*, September 26, 2002, in *Alexander's* 7, no. 20 (October 15, 2002). Canada's most recent involvement is reported in "CNOOC Signs Production Sharing Contracts with Husky," *Xinhua*, September 24, 2002, in *Alexander's* 7, no. 20 (October 15, 2002).

⁶¹ Xie Ye, "New Oil Regulations Allow Sinopec to Cooperate with Foreign Companies," *China Daily*, October 16, 2001, in FBIS-CPP20011016000040. These revised regulations were codified into an amended Statute on the Exploitation of Land Oil Resources by Foreign Enterprises on September 23, 2001 ("Zhu Ronghi Signs Amendment to PRC Statute," *Xinhua*, October 10, 2001, in FBIS-CPP20011010000184).

⁶² "Minister on Use of Foreign Capital in 2002," *Xinhua*, December 27, 2001, in FBIS-CPP20011227000151.

⁶³ "China Seeks Foreign Investors for Shanghai Gas Grid," Reuters, in *Alexander's* 7, no. 4 (February 21, 2002). Also see "Huaneng Power to Set Up Two Gas-Fuelled Power Plants," Interfax Information Services, August 28, 2002, in *Alexander's* 7, no. 18 (September 19, 2002): one of these plants is planned for Shanghai, the other for Nanjing.

⁶⁴ Wang Ling, "Zhejiang Province's Hangzhou City Readies Itself for Switch to Natural Gas," *China Daily*, March 28, 2002, in FBIS-CPP20020328000048; "Zhejiang to Build Natural Gas-Fuelled Power Plant," Interfax Information Services, September 28, 2002, in *Alexander's* 7, no. 19 (October 1, 2002), links this project—projected to begin construction in 2003—to the West-East natural gas pipeline.

⁶⁵ "Gas Supplier in Shanghai Competes with Rivals in West China," *Xinhua*, August 17, 2002, in FBIS-CPP20020817000082.

⁶⁶ CNPC has been active in Azerbaijan; see "Chinese Oil Major Eyes Share in Azerbaijani Onshore Field," *Baku MPA*, January 28, 2003, in FBIS-CEP200301228000065.

⁶⁷ Jiang Zemin, quoted in "Summit Meeting Launches Shanghai Cooperation Organization," *People's Daily*, June 15, 2001, accessed at <www.english.peopledaily.com.cn/200106/15/eng20010615_727040.html>.

⁶⁸ Anthony Miccarelli, "China's Energy Future: The 10th Five-Year Plan (2001–2005)," USCINCPAC Virtual Information Center (VIC) Honolulu, accessed at <www.cschuster@vic-info.org> (August 25, 2001), 13.

⁶⁹ "China Set to Invest $300 Million in Kyrgyzstan's Oil Sector," Kabar News Agency, June 26, 2002, in FBIS-CEP20020626000164.

⁷⁰ For instance, see Xie Ye, "State-Owned CNPC Buys Two Overseas Oilfields," *China Daily*, January 25, 2002, in FBIS-CPP20020125000026; "Chinese Firms Interested in Pakistan Oil and Gas Exploration," *Asia Pulse*, in *Alexander's* 7, no. 2 (January 23, 2002); "China to Help Uzbekistani Oil, Gas Firm," *Novosti Nedeli*, November 16, 2001, in FBIS-CEP20011118000052; "U.S., China and Iran Compete for Afghan Oil and Gas Pipelines," *Asia Times*, in *Alexander's* 7, no. 4 (February 21, 2002).

⁷¹ Cited in David Lague, "China: The Quest for Energy to Grow," *Far Eastern Economic Review* (June 20, 2002), accessed at <www.feer.com>. This number is probably exaggerated but represents China's clear interest in making serious inroads into Central Asian petroleum resources.

[72] "China Ready to Import Up to 50 Million Tonnes of Kazakh Oil a Year," *Almaty Interfax-Kazakhstan,* January 29, 2003, in FBIS-CEP20030129000590, is not a disinterested source, but large oil sales to China certainly may be in the offing.

[73] "Kazakhstan-Western China Oil Pipeline Said Medium-Term Prospect," *Kazakhstan Today* (Almaty), September 11, 2001, in FBIS-CEP20010912000202.

[74] See, for instance, "Chinese Interested in Kazakhstani Oil Projects," Caspian News Agency, and "Kazakhstani and Chinese Institutes to Cooperate in Oil and Gas Sector," both in *Alexander's* 6, no. 20 (October 24, 2001).

[75] Cited in "China to Build Oil Pipeline from Kazakhstan if Caspian Reserves Proved," *Almaty Interfax-Kazakhstan,* April 16 2002, in FBIS-CEP20020416000327. One interesting factor in the issue of recovering Caspian Sea oil is the apparent agreement between Iranian and U.S. representatives to favor a pipeline to Iran instead of the Chinese project ("Kazakhstani Paper Views Possible Export Routes for Caspian Oil," *Kazakhstanskaya Pravda,* April 10, 2002, in FBIS-CEP20020410000253).

[76] This discussion draws on Rick Parker, "China's Energy Strategy: Central Asian-Caspian Sea Oil Initiatives," VIC (January 22, 2001).

[77] Greg Austin, *China's Ocean Frontier: International Law, Military Force, and National Development* (Canberra: Allen and Unwin, 1998), 262, notes that this area contains "two of the biggest PRC offshore gas fields in the South China Sea," Yacheng 13–1 and Dongfang 1–1. The best general account of China's activities in the South China Sea is Mark J. Valencia, *China and the South China Sea Disputes,* Adelphi Paper 228 (London: Institute for International Strategic Studies, 1995).

[78] China and Vietnam continue to contest various prospecting blocks in the South China Sea, although ongoing talks have reduced the number of incidents between the two nations. See Valencia for an account of these conflicting claims; one indicator of continued Chinese interest is "CNOOC Invites Foreign Firms in South China Sea Projects," *People's Daily,* September 25, 2002, in *Alexander's* 7, no. 20 (October 15, 2002).

[79] "South China Sea Region," DOE-EIA, August 1998, 2.

[80] "South China Sea Oil Fields Produced Over 50 mm Tons in Eight Years," *Alexander's* 3, no. 20 (October 27, 1998).

[81] Ibid., 3. The author interviewed two analysts in November 1999 who, using similar geological data, came to opposite conclusions about the presence or absence of significant petroleum reserves in the Spratlys.

[82] U.S. Geodetic Survey (USGS), in USCINCPAC memo, September 15, 1999. One optimistic estimate of South China Sea petroleum reserves, 55 billion tons, is in *Xinhua,* September 5, 1994, in FBIS-CHI-94-172. Also see Michael Leifer, "Chinese Economic Reform and Security Policy: The South China Sea Connection," *Survival* 37, no. 2 (Summer 1995), 44; Bruce Blanche and Jean Blanche, "Oil and Regional Stability in the South China Sea," *Jane's Intelligence Review* 7, no. 11 (November 1995), 513. The USGS estimate for South China Sea natural gas is 266 trillion cubic feet.

[83] "Shell Invests $3 Billion for Gas Development in PRC," *Xinhua,* September 29, 1999, in FBIS-CPP19990920999635, reported that "the South China Sea Petrochemical Project, co-funded by Shell Group and several large Chinese enterprises . . . has a total investment of 4.5 billion U.S. dollars."

[84] Brief descriptions of some of this technology is in "Lufeng 22–1," BP Amoco announcement October 18, 1999, 1–3, accessed at <www.offshoretechnology.com/projects/lufeng/index>. This field is a joint venture between CNOOC and Staatoil of Norway.

[85] Author's conversations with PRC Embassy (Washington) international lawyer and with senior People's Liberation Army Navy officers at the PLA National Defense University and China's Naval Research Institute (2000–2002).

[86] While such estimates probably reflect political wishes as well as geophysical facts, this number is in "China Kicks Off Huge Offshore Oil, Gas Project," *Xinhua,* June 19, 2002, in FBIS-CPP20020619000059.

[87] "Ocean Petroleum Lays Exploration Emphasis on Bohai Bay," *AsiaPort,* April 8, 2002, in *Alexander's* 7, no. 9 (May 3, 2002), notes that half of these wells would be drilled by "partners." "In Brief," *China Daily,* May 22, 2002, in FBIS-CPP20020522000054, noted that CNOOC and several foreign companies were some of these "partners"; also see "Ultra Petroleum Signs Bohai Bay Agreement with CNOOC," Ultra Petroleum announcement in *Alexander's* 7, no. 19 (October 1, 2002), and "Shell to Invest $400 Million in China's Bohai," *Neftegaz,* July 24, 2002, in *Alexander's* 7, no. 16 (August 23, 2002).

⁸⁸ "China, Vietnam Initial Agreements on Beibu Gulf Demarcation," *People's Daily*, December 25, 2000, accessed at <www.english.peopledaily.com.cn/200012/25/eng20001225_58714.html>.

⁸⁹ See "Eight Major Highways, 3,000 km of Railways to Speed Development," *China Daily* (Hong Kong), May 23, 2001, in FBIS-CPP20010523000067, for a report of the "nearly 20 foreign energy companies" bidding to participate in pipeline projects linking China's far west with its coastal region; "Kerr-McGee and Partners Strike More Oil in Bohai Bay," Reuters, in *Alexander's* 7, no. 1 (January 9, 2002). "Ultra Petroleum Announces Oil Field Discovery in Bohai Bay," Ultra Petroleum announcement, in *Alexander's* 7, no. 1 (January 9, 2002), describes a company "focused on developing its long life natural gas reserves in the Green River Basin of Wyoming, and oil reserves in Bohai Bay." The British company BP Amoco has been especially active; see "World Oil Giant BP to Invest More in Chongqing," *Xinhua*, October 31, 2001, in FBIS-CPP20011031000043; Xie Ye, "China's Largest Petrochemical Firm Enters Joint-Venture with UK's BP Oil," *China Daily*, December 11, 2001, in FBIS-CPP20011211000022; Xie Ye, "Sinopec, UK's BP to Launch Joint Venture," *China Daily*, December 12, 2001, in FBIS-CPP20011212000041; and especially "World Oil Giant BP to Invest $5 Billion in China," *Xinhua*, November 11, 2001, in FBIS-CPP20011111000060. The author suspects the profit motive outweighs Chinese companies' concern with "owning" at least 51 percent of a given joint venture. Jing Ji, in "Sinopec, ExxonMobil Sign Framework Agreement to Strengthen Strategic Alliance," *China Daily*, October 23, 2002, 2, in FBIS-CPP20021023000044, for instance, reports that ExxonMobil and Aramco would each own 25 percent of "multi-billion-dollar petrochemical and oil product joint ventures" in Fujian and Guangdong provinces, while the remaining 50 percent would be split between Sinopec and "the Fujian provincial government."

⁹⁰ Quoted in "Zhu Rongji on China Opening West-East Gas Project to Foreign Cooperation," *Xinhua*, July 4, 2002, in FBIS-CPP20020704000114.

⁹¹ The project is described in "Gas Pipeline Runs Over 4,000 KM Across China," *Renmin Ribao*, July 5, 2002, in FBIS-CPP20020705000036; the UNDP study in "UNDP and Shell to Assess Social Impact of China Pipeline Project," *UN News*, April 17, 2002, in *Alexander's* 7, no. 9 (May 3, 2002), is relying on "government organizations, academics, and independent international social experts."

⁹² Oil production is reported in "Another of China's Xinjiang Fields Passes Million-ton Oil Output," *OGI*, in *Alexander's* 8, no. 5 (March 6, 2003). See, for instance, "PetroChina Believes Domestic Natural Gas Reserves are Sufficient," *China Daily*, July 17, 2002, in *Alexander's* 7, no. 16 (August 23, 2002) and "China Negotiating With British Petroleum for Fujian Gas Deal," *China Daily* (Business Weekly Supplement), August 20, 2002, in FBIS-CPP20020821000018.

⁹³ "China, Russia to Jointly Develop Oil Fields in Siberia," *Xinhua*, September 27, 2001, in FBIS-CPP20010927000088. The success of these talks was reported in Xie Ye, "Talks on Oil Project Underway," *China Daily*, September 24, 2001, in FBIS-CPP20010924000017, and "Russia and China Begin Work on Oil Pipeline Feasibility Study," *Vremya Novostei*, in *Alexander's* 6, no. 20 (October 24, 2001), which reported the pipeline's initial (2005) capacity as 20 million tons, increasing to 30 million tons by 2010.

⁹⁴ "Joint Declaration of the PRC and the Russian Federation," *Xinhua*, December 2, 2002, in FBIS-CPP20021202000186.

⁹⁵ "China to Close More Small Oil Refineries," AFP, February 28, 2000, quoted in *Alexander's* 5, no. 7 (April 27, 2000).

⁹⁶ Figures cited, but their sources not listed, in Chang, 226.

⁹⁷ Peng Sen, Vice Minister of the Economic Restructuring Office of the State Council, quoted in "China Needs to Establish Downstream Gas Sector Regulatory Framework," *Xinhua*, April 10, 2002, in *Alexander's* 7, no. 9 (May 3, 2002), and Chen Li, a "senior official" in the same office, in "Chinese Experts Call for Establishment of Gas Market Watchdog," *China Daily*, April 9, 2002, in *Alexander's* 7, no. 9 (May 3, 2002). Also see "The Bulls Are Back," *Petroleum Economist* 67, no. 11 (November 2000), 38–39.

⁹⁸ The 8 percent figure is cited in "China's Modernization May Be Slowed Down by Oil Shortage," *People's Daily*, in *Alexander's* 6, no. 15 (August 14, 2001); Mai Tian, "Sinopec, PetroChina Reach Agreement on Pipeline Project," *China Daily*, October 6, 2001, in FBIS-CPP20011006000032, reports the goal as 10 percent.

⁹⁹ Author's interview of geologist, at the East-West Center, Honolulu (November 1999).

¹⁰⁰ See, for instance, "Sinopec Looks to Boost its Ability in the Natural Gas Arena," *People's Daily*, in *Alexander's* 6, no. 18 (September 25, 2001); "Chinese Cities to Benefit From West-East Natural Gas Transmission Project," *Star*, in *Alexander's* 6, no. 20 (October 24, 2001). Liquefication of natural gas is

required for ease of shipping and because current technology limits the distance over which unlique-
fied gas can be piped.

[101] "China to Require Natural Gas Imports From 2005," *Interfax Information Services, B.V.*, January 29, 2003, in *Alexander's* 8, no. 4 (February 20, 2003).

[102] "Gas Reserves of Changqing Oilfield Equal 1 Trillion Cubic Meters," *Xinhua*, November 25,
2001, in FBIS-CPP20011125000030.

[103] Announced by Qiu Zhongjian, director-general of the China Petroleum Society, in "China
Expects to Discover More Reserves of Natural Gas in Next 10 Years," *Xinhua*, September 6, 2001, in
FBIS-CPP20010906000097, who listed three primary areas of expansion, all in western China: the
Tarim Basin, Ordos Zone, and Qaidam Zone; the Sichuan Province also is promising.

[104] Government report cited in "China Boosts Natural Gas Industry," *Asia Pulse*, in *Alexander's* 7,
no. 2 (January 28, 2002).

[105] Gaye Christoffersen, "Prospects and Problems for Northeast Asian Energy Cooperation,"
IREX/Huang Hsing Foundation Hsueh Chun-tu Lecture Series, accessed at <asia-1@info.irex.org>.

[106] See Mark J. Valencia, "Energy and Insecurity in Asia," *Survival* 39, no. 3 (Autumn 1997),
85–106, for a discussion of some of these schemes, especially the map on page 88. "ASEAN Plans Pan-
Asian Pipeline," Dow Jones, in *Alexander's* 6, no. 10 (June 1, 2001), discusses a proposal to build a
pipeline system that would also include India.

[107] See Zhu Rongji, "Government Work Report at the Opening of the Fifth Session of the Ninth
National People's Congress," March 5, 2002, in FBIS-CPP20020305000078.

[108] "PRC's Gas Pipeline Project to Begin 'In a Short Time'," *Xinhua*, September 6, 2001, in FBIS-
CPP 20010906000154, cites a $4.8 billion estimate, while "China to Build E-Pipeline for Natural Gas
Transmission Project," *Xinhua*, September 21, 2001, in FBIS-CPP20010921000086, cites $5.3 billion;
"Another Gas Field Found in China's Far Northwest," *AsiaPort*, in *Alexander's* 6, no. 20 (October 24,
2001), gives a $7 billion dollar estimate. Different reports give 2003 or 2004 as the completion date,
although Liang Yu simply reports that construction has "been put off" ("East-West Gas Pipeline Project Delayed," *China Daily* (Business Weekly Supplement), October 9, 2001, in FBIS-
CPP20011010000007).

[109] "Nissho Iwai Withdraws from China Gas Project," *Kyodo News*, in *Alexander's* 6, no. 20 (October 24, 2001), reported "similar moves by five other Japanese trading houses and means no Japanese
businesses will be involved in the pipeline project."

[110] "PetroChina and Shell Group Sign Gas Pipeline Interim Pact," *Platts*, in *Alexander's* 7, no. 4
(February 21, 2002). Also see "PetroChina and Shell/Gazprom Sign Gas Trunkline Deal," Reuters, in
Alexander's 7, no. 2 (January 23, 2002), for the report that BP was awarded a subsidiary contract "to
build China's first LNG terminal in Guangdong." This would expand Shell's already impressive investment of more than $5 billion in China's energy sector ("Shell's Investment in China," *Xinhua*, March
26, 2002, in FBIS-CPP20020326000166).

[111] "900-KM Section of West-East Gas Pipeline Completed at the End of August," *Xinhua*, September 23, 2002, in VIC, "Asia-Pacific Daily News Summary" (September 25, 2002). Completion of
this section is not supported by other reporting, but "Gansu to Start Laying Gas Pipeline," *Xinhua*,
February 25, 2003, in FBIS-CPP20030225000145, reported a March 1 start date for a section of the
pipeline in Gansu Province.

[112] "Zhejiang Province Builds Facilities to Increase Use of Natural Gas," *Xinhua*, January 1, 2003,
in *Alexander's* 8, no. 2 (January 24, 2003).

[113] "Shell to Operate China's Leading Gas Field," Reuters, May 31, 2002, in *Alexander's* 7, no. 13
(June 27, 2002).

[114] "Xinjiang Says to Supply Gas to Shanghai for 30 Years," *Xinhua*, September 24, 2001, in FBIS-
CPP20010924000013, includes the estimate by an unidentified analyst that the Tarim Basin will "be
able to provide gas to energy-deficient Shanghai for more than 30 years." This estimate is suspiciously
optimistic, however, since it is attributed to the "general manager of the Talimu Oil Field Company
(Liao Yongyuan)" quoted in "Talimu Oil Field Could Supply Natural Gas to Shanghai for 30 Years,"
AsiaPort, in *Alexander's* 6, no. 20 (October 24, 2001).

[115] "Section of West-East Natural Gas Transmission Project Launched," *AsiaPort*, in *Alexander's*
7, no. 5 (March 6, 2002), reports the beginning of a section of this project. Other associated construction efforts are reported in "China Starts Building Gas Pipeline from Xinjiang to Shanghai," Reuters,
in *Alexander's* 7, no. 5 (March 6, 2002).

[116] "Shanghai Would Promote Employment of Natural Gas," *AsiaPort*, in *Alexander's* 7, no. 6 (March 21, 2002).

[117] "Russian GazProm Gas Company Opens Office in Beijing," ITAR–TASS, December 26, 2001, in FBIS-CEP20011226000350. "Russia and China to Set Up Joint Gas Project Enterprises," ITAR-TASS, in *Alexander's* 6, no. 18 (September 25, 2001), reported that this agreement included more far-ranging Russian participation in China's developing natural gas infrastructure, a "full-scale strategic partnership."

[118] "PetroChina Signs Gas Deal with Chinese Cities," Reuters, in *Alexander's* 7, no. 4 (February 21, 2002), reports "no date has yet been set for the start of construction."

[119] "Important Gas Reserves Discovered in Sichuan Province," *Xinhua* (January 8, 2003), in FBIS-CPP20030108000138, reports that fields containing 150 billion cubic meters have now been discovered in Sichuan. Also see "Collapse of Enron Complicates China Pipeline Project," Reuters, in *Alexander's* 7, no. 3 (February 6, 2002).

[120] Xie Ye, "Gas Projects to Power Country's Market Growth," *China Daily*, March 20, 2002, in FBIS-CPP20020320000039.

[121] "China Boosts Natural Gas Industry"; "CNOOC to Build China's Second-Longest Gas Pipeline," *Neftegaz.RU*, December 23, 2002, in *Alexander's* 8, no. 2 (January 24, 2003), reports plans to build "China's second-longest natural gas pipeline" from Jiangsu and Zhejiang with Guandong, although with a start date of 2005 at the earliest.

[122] Zhao Renfeng, "Indonesia Confident of Success in Guangdong Natural Gas Bid," *China Daily* (Business Weekly Supplement), April 2, 2002, in FBIS-CPP20020402000057.

[123] Xie Ye, "Bidding Under Way for First Chinese LNG Project," *China Daily*, November 10, 2001, in FBIS-CPP20011110000056. The BBC reported in December 2001 that the project's feasibility study was scheduled for completion in June 2002, with construction to begin by the end of the year; "China Opens Bidding for Guangdong LNG Supply Contract," in *Alexander's* 6, no. 24 (December 19, 2001). Natural gas is normally measured in cubic feet or cubic meters; LNG is usually measured in tons.

[124] Xie Ye, "Hong Kong to Buy One Third of Imported LNG," *China Daily*, August 13, 2002, in FBIS-CPP20020813000013.

[125] "Australia Wins Bidding to Supply LNG to China," *Xinhua*, August 8, 2002, in FBIS-CPP20020808000021.

[126] "China's Offshore Oil Giant CNOOC Unveils 2002 Development Strategy," *Xinhua*, January 18, 2002, in FBIS-CPP20020118000098.

[127] "CNOOC Begins Gas Terminal Construction in Hainan," Reuters, in *Alexander's* 7, no. 3 (February 6, 2002).

[128] In "CNOOC and Sinopec to Develop Gas Field in East China Sea," *Gulf News Online*, in *Alexander's* 7, no. 2 (January 23, 2002), the Chinese companies threaten to develop the offshore fields "with or without foreign help," and while a March 20, 2002, article reports that "by 2004, natural gas from the Chunxiao Gas Field Group will be piped ashore," it is doubtful that Shanghai will receive significant amounts of natural gas from this project without foreign participation, given Chinese companies' dependence on foreign investment and participation in the offshore energy sector. (See "China Taps Natural Gas Reserves in East China Sea," *Xinhua*, March 20, 2002, in FBIS-CPP20020320000165.)

[129] See "Siemens Aims to Set Up Joint Venture with Shanghai Electric," Reuters, in *Alexander's* 7, no. 3 (February 6, 2002).

[130] Xie Ye, "CNOOC Seeks to Buy Overseas Gas Fields," *China Daily* (Business Weekly Supplement) in FBIS-CPP20011127000104. Probably more authoritative is "Indonesian and Chinese Firms Sign Major Energy Deals," Associated Press, September 26, 2002, in *Alexander's* 7, no. 20 (October 15, 2002), which quotes Indonesian and Chinese officials reporting that their national companies signed six memoranda of understanding "worth hundreds of millions of dollars to cooperate in [Indonesia's] oil, mining and power sectors." Details are in Johannes Simbolon, "RI, China Firms Sign Major Deals on Energy," *The Jakarta Post*, September 26, 2002, in FBIS-SEP20020926000002.

[131] "Iran to Allow Chinese Oil Giant to Explore Near Tehran," Iranian News Agency, January 13, 2001, in FBIS-IAP20010113000002. "Iran is Working to Attract Investment from India and China," AFP, in *Alexander's* 7, no. 23 (November 27, 2002), also notes that "Iran holds an estimated 18 percent of the world's natural gas reserves."

[132] Han Rongliang, "China to Build Three Channels for West-East Electricity Transmission," *Renmin Ribao Radio*, March 20, 2002, in FBIS-CPP20020320000090.

[133] "More Schemes Unveiled to Tap Yangtze Hydropower Resources," *China Daily*, February 12, 2003, in FBIS-CPP20030213000014.

[134] "China to Make Investment for Three Gorges Power Transmission," *Xinhua*, October 31, 2002, in *Alexander's* 7, no. 23 (November 27, 2002).

[135] "World's Largest Hydropower Station Launched in China," *Renmin Ribao*, October 25, 2002, in FBIS-CPP20021025000062.

[136] Manning, 97–98.

[137] "China-Built Nuclear Generator Unit Operational," *Xinhua*, August 16, 2002, in FBIS-CPP20020816000085; "China's 1st Heavy Water Generator Starts Working," *China Daily*, November 20, 2002.

[138] "Winshan Nuclear Power Plant Begins 2d Phase Operations," *Xinhua*, February 6, 2002, in FBIS-CPP20010106000095 and "China-Made Nuclear Plant Runs for Commercial Use," *Xinhua*, April 19, 2002, in FBIS-CPP20020419000041; "Three Nuke Power Plants Ready for Use this Year," *People's Daily*, January 31, 2002, accessed at <www.english.peopledaily.com.cn/200201/>.

[139] Li Nanling, "Lingao Nuclear Power Station Loads Fuel, Begins Trial Run," *Xinhua* (Hong Kong), December 9, 2002, in FBIS-CPP20011209000050.

[140] Xie Ye, "China to Build New Nuclear Power Plants in Coastal Provinces," *China Daily*, March 13, 2001, in FBIS-CPP20010313000061, and "More Nuke Power on the Way," *China Daily*, December 17, 2001, in FBIS-CPP20011217000025.

[141] Xie Ye, "New Nuke Power Plant Planned," *China Daily*, January 11, 2003, in FBIS-CPP20030111000023, reporting that Zhejiang Province had applied to Beijing to build a "U.S. $2 billion nuclear power plant" in 2004.

[142] "China's Nuclear Power Capacity to Keep Rising," *Xinhua*, October 23, 2002, in FBIS-CPP20021023000137. "Construction of Nuclear Power Plant Completed in Shenzhen," *Xinhua*, January 8, 2003, in FBIS-CPP20030108000065, reports completion of Guangdong Province's second 2-million-kW nuclear power plant.

[143] "China's 1st Heavy Water Generator Starts Working," *China Daily*, November 20, 2002, also notes that the power company is negotiating with the government for tax breaks based on the fact "given nuclear power is cleaner energy compared with thermal power." "State of the Art Reactor Faces an Uncertain Future," AFP, in the *South China Morning Post*, November 21, 2002, reported (probably erroneously) that the third plant at Qinshan began producing power in November and expanded on the company's appeal to the government and difficulty in repaying its construction loans.

[144] Liu Jiang and Han Song, "China's New-Generation Nuclear Reactor Successfully Connects to Power Grid in Full Capacity for Generating Electricity in Beijing," *Xinhua*, March 1, 2003, in FBIS-CPP20030301000081, is mistitled, since the plant is intended initially for experimentation and personnel training.

[145] "Russia to Continue Building Nuclear Power Plants Abroad," ITAR–TASS, March 27, 2002, in FBIS-CEP20020327000315. Also see, for instance, "Russians to Start Supplying Equipment for Chinese Nuclear Power Plant," Russian Information Agency (Moscow), in FBIS-CEP20010918000085; Andrei Kirilov, "Specialists Install Nuclear Reactor in China," ITAR–TASS, April 20, 2002, in FBIS-CEP20020420000045; finally, Russia announced it would bid on additional nuclear power plant construction in China in "Russia's Atomic Energy Minister Visits China," ITAR–TASS, July 8, 2002, in FBIS-CEP20020708000036.

[146] Manning, 98–100.

[147] "Analysts Say China Making 'Immense' Progress Producing Nuclear Power Equipment," *Xinhua*, March 30, 2002, in FBIS-CPP20020330000089, reports successful manufacture of "reactor components" for the second phase of the Qinshan power plant, but the very celebratory tenor of this report bespeaks the nascent nature of this industrial capability in China—especially in view of the more authoritative reports that this equipment was obtained from Canada.

[148] "Fujian Province to Build Wind-Generated Power Plants," *Xinhua*, in *Alexander's* 8, no. 4 (February 20, 2003).

[149] "CLP to Build China's Largest Wind Power Plant in Guangdong," *SinoCast*, January 15, 2003, in *Alexander's* 8, no. 2 (February 6, 2003).

[150] "China to Halve Value-Added Levy on Wind Power Generation," *Asia Pulse*, in *Alexander's* 7, no. 11 (May 29, 2002).

[151] "China Has Bright Prospects for Biogas Power Plants," *Asia Pulse*, in *Alexander's* 6, no. 24 (December 19, 2001), reported that city garbage was being burned as an energy source in Hangzhou

and Guangzhou, as did "Northeast China's Harbin Builds Waste-Burning Power Plant," *Xinhua*, April 7, 2002, in FBIS-CPP20020407000064, with Nanjing, Shanghai, Beijing, Shenzhen, and Beihai planning to follow suit.

[152] "China Promoting Ethanol-Based Fuel on Pilot Basis in Five Cities," *Xinhua*, June 17, 2002, in FBIS-CPP20020617000134.

[153] Qin Chuan, "Landfill Gas Plan Powers Future Free of Pollution," *China Daily*, October 24, 2002, in FBIS-CPP200210240000042, describes a seemingly far-fetched fuel processing scheme but represents Beijing's determination to find alternative energy sources.

[154] "State Power Corporation of China Invested $18 Billion in Upgrade of Power Grids," *China Daily*, April 8, 2002, in *Alexander's* 7, no. 9 (May 3, 2002).

[155] "Guangxi Region to Invest in New Power Stations," *Xinhua*, August 27, 2002, in *Alexander's* 7, no. 18 (September 19, 2002). The Yunnan plan is in Miccarelli, "China's Energy Future," 16. Also see "Shanghai to Upgrade Power Grid," *Xinhua*, February 25, 2003, in FBIS-CPP20030225000229, for that municipality's plan to invest $445 million to upgrade its power grid in 2003; and "South China Province to Regulate Large-Scale Electricity Users," *Xinhua*, in *Alexander's* 8, no. 5 (March 6, 2003), for Guandong Province's plan to improve power grid management. "Shenzen to Invest $325.3 Million in Electricity Grids," *Xinhua*, January 23, 2003, in FBIS-CPP20030122000232, provides details of Shenzhen's plan.

[156] See Fu Jing, "State Funds Ready to Light Up Rural Towns," *China Daily*, February 11, 2003, in FBIS-CPP20030211000035.

[157] "China Reports on Receipt of New Funds in Energy Sector," *The PMA Online Power Report*, August 6, 2002, in *Alexander's* 7, no. 16 (August 23, 2002), reported that the power industry received $11.9 billion and also noted that the coal sector received the smallest amount ($8.3 million) of new investment funds.

[158] "Tibet to Exploit Hydro, Solar Energy," *China Daily*, May 23, 2001, in FBIS-CPP20010523000041, and "Tibet Sets Development Strategy for Power Sector," *Alexander's* 6, no. 24 (December 19, 2001); other projects are discussed in "Most Power Projects in Operation in China's Guangxi Region," "Inner Mongolia and Beijing to Jointly Develop Power Plant Project," and "China Lists Planned Power Projects," all in *Asia Pulse* (October 24, 2001). The claim about generating power for this hydropower system is in "China to Build Four Major Hydropower Plants in Southwest," *Xinhua*, December 7, 2001, in FBIS-CPP20011207000080; these plants are not scheduled for construction until 2005 and 2006. "Hydropower Plant Operational in Yunnan," *Xinhua*, December 14, 2001, in FBIS-CPP20011214000121. Also see "Eight Power Stations to be Built on Lancang River," *People's Daily*, January 20, 2002, accessed at <www.english.peopledaily.com.cn>, and "PRC Touts Lancang River as Energy Base," *Xinhua*, January 31, 2002, in FBIS-CPP20020131000071, which also notes that Thailand and Laos will buy some of this hydropower.

[159] "China Set to Improve Investment Environment for Power Industry," *Xinhua*, June 28, 2001; Noah Smith, "Focus on China's Energy Restructuring Plan," *Kyodo News*, August 7, 2001, in FBIS-JPP20010807000143. The Gansu plan includes an upgrade to the Lanzhou gas-fuelled power plant and three power grid improvements ("Gansu Power Receives Approval for Four Power Projects," *Interfax Information Services*, July 15, 2002, in *Alexander's* 7, no. 16 (August 23, 2002).

[160] The most concise description of the new distribution infrastructure is in "Profile of China's Electric Power Industry," *Asia Pulse*, January 6, 2003, in *Alexander's* 8, no. 2 (January 24, 2003). Also see Li Heng, "China Unveils Plan for Power Sector Regrouping," *People's Daily*, October 23, 2002, in FBIS-CPP20021023000080; "China to Split State Power Corp," *Reuters*, in *Alexander's* 7, no. 5 (March 6, 2002); Pamela Pun, "Beijing to Set Up Regulatory Body to Draft Policies, Regulate Power Industry," *Hong Kong iMail*, April 19, 2002, in FBIS-CPP20020419000063. Wang Xiangwei, "Top Mainland Official Flees to Avoid Arrest," *South China Morning Post*, October 16, 2002, reported that Gao Yan, "the former president and chief executive" of China's State Power Corporation, fled China to avoid prosecution for corruption.

[161] "China to Invest 360 Billion Yuan in Construction of Power Grids," *Xinhua*, in *Alexander's* 7, no. 6 (March 21, 2002). "China's Shandong Province to Invest in Power Projects," *Asia Pulse*, in *Alexander's* 7, no. 3 (February 6, 2002) notes that $1.72 billion will be invested in power projects in this province in 2002.

[162] "China Puts End to Monopoly on Power Industry," *Xinhua*, December 30, 2002, in *Alexander's* 8, no. 2 (January 24, 2003).

[163] World Health Organization (1999), accessed at <www.angelfire.com/mi2/pnginc/ccten-worst.htm>.

[164] "China Issues Latest Technology Policy to Promote Clean Energy," Interfax Information Services, in *Alexander's* 7, no. 16 (August 23, 2002).

[165] "China Forecasts on Installed Power Capacity," Interfax Information Services, May 27, 2002, in *Alexander's* 7, no. 12 (June 13, 2002).

[166] "Beijing Paper Examines How New Coal-Burning Boilers to Create Cleaner Air," *Xinhua*, October 29, 2002, in FBIS-CPP20021029000109.

[167] Tian Xiuzhen, "City Gets Tougher on Emissions," *China Daily*, February 26, 2003, in FBIS-CPP20030226000087, also notes the World Expo planned for Shanghai in 2010.

[168] "China Implements 'Green Project' for Successful Olympics," *Xinhua*, April 11, 2002, in FBIS-CPP20020411000144.

[169] "Beijing to Boost Geothermal Energy Resources," Interfax Information Services, in *Alexander's* 7, no. 23 (November 27, 2002).

[170] See "Seminar Held on 'Green Light Program' to Protect Environment, Save Energy," *Xinhua*, September 18, 2001, in FBIS-CPP20010919000001; "Beijing to Develop Pollution-Free Energy, Cut Coal Consumption," *Xinhua*, February 24, 2002, in FBIS-CPP20020224000013.

[171] Tian Xiuzhen, "Pollution, Energy Top Concerns for Shanghai," *China Daily*, November 8, 2000, in FBIS-CPP20002208000032.

[172] "China Begins Construction on 4,200-Kilometer Long Gas Pipeline Project," *Xinhua*, July 4, 2002, in FBIS-CPP20020704000141, is too optimistic, but shows that environmental improvement is becoming a more prominent concern in China's energy infrastructure plans.

[173] Cited in Barbara A. Finamore, "Taming the Dragon Heads: Controlling Air Emissions from Power Plants in China" (Taipei: Natural Resources Defense Council, June 2000).

[174] "More Small, Eco-Friendly Hydropower Stations Built in Rural China," *Xinhua*, February 1, 2002, in FBIS-CPP20020201000069.

[175] Liang Chao, "U.S. $3.6 Billion Pledged for Water Projects," *China Daily*, February 26, 2003, in FBIS-CPP20030226000092.

[176] Luan Shengji and Hong Yang, "Water Security Problem in the 21st Century," *Keji Ribao* [*Science and Technology Daily*], March 9, 1998, in FBIS-CHI-98-217.

[177] "Northern PRC Cities Facing 'Serious' Water Shortage," *Xinhua*, April 2, 2002, in FBIS-CPP20020402000170; Zhu Ronghji, quoted in "Chinese Premier Unveils 'Green Plan,'" March 5, 2001, accessed at <www.cnn.com./news>.

[178] Xie Ye, "Firms Thirst After Canal Project," *China Daily* (Business Weekly Supplement), November 19, 2000, in FBIS-CPP20001119000015. The desalination scheme is discussed in "China Develops Nuclear Powered Heating and Desalination," *Xinhua*, June 20, 2002. "West Route of Water Diversion Project to Start in 2010," *Xinhua*, August 18, 2002, in FBIS-CPP20020818000047, provides some additional but still sketchy details, including an estimated cost of approximately $41 billion for two of the three routes; Miccarelli, "China's 'Three Canals': The Impact of China's Water Diversion Project," VIC (May 9, 2001), gives $30 billion as the cost. Both of these figures are almost certain to be exceeded in such an unprecedented project.

[179] "Lancang-Mekong River Opens to Navigation in China, Laos, Burma, Thailand," *Xinhua*, June 26, 2001, in CPP20010626000018; "China to Help Myanmar, Laos Dredge Mekong River," *Xinhua*, June 28, 2001, in FBIS-CPP20010628000151; and "Project to Improve Navigation Route Through China, Burma, Laos, Thailand, Cambodia, Vietnam," *Xinhua*, March 30, 2002, in FBIS-CPP20020330000095, all reflect both the sensitivity of plans to affect the course of these rivers, and Beijing's plan to defuse opposition.

[180] "China to Provide Information on Mekong River to Downstream States," *Xinhua*, April 1, 2002, in FBIS-CPP20020401000024; "Six Mekong Countries Sign Regional Power Trade Accord," *Xinhua*, November 3, 2002, in FBIS-CPP20021103000055.

[181] "China to Offer Hydrological Data to Mekong River Commission," *Xinhua*, June 12, 2002, in FBIS-CPP20020612000068; "GMS Cooperation Meeting Opens in Kunming," *Xinhua*, June 8, 2002, in FBIS-CPP20020608000071.

[182] "Work Begins on First Overseas-Funded Hydropower Station on Huang He," *Xinhua*, September 25, 2002, in FBIS-CPP20020925000218. "CNPC and Power Technology Forge Ahead on Targets," *PWTC/APMT*, June 6, 2002, in *Alexander's* 7, no. 13 (June 27, 2002), reports China's attempts to secure advanced welding technologies to improve pipeline safety and durability.

[183] Li Dadong, director of the China Petrochemical Industry Scientific Research Institute, quoted in "China to Vigorously Develop its Clean Fuel Technology," *AsiaPort*, in *Alexander's* 6, no. 24 (December 19, 2001), discusses alternate automotive fuels; "Many Chinese Cities Generate Power by Rubbish and Marsh Gas," *AsiaPort*, in *Alexander's* 6, no. 20 (October 24, 2001); "Western China Area May Also Use Renewable and Green Energy," *Asia Pulse*, in *Alexander's* 6, no. 18 (September 25, 2001); "Hebei Province Farmers Benefit From Straw Gasified Power," *Xinhua*, March 12, 2001, in FBIS-CPP20010312000163; "China Develops Green Hydrogen Energy," *Xinhua*, January 17, 2002, in FBIS-CPP20020117000164, discusses various projects—especially a joint venture with a Canadian company—to develop fuel cell technology, which "will not produce any pollutant." Also see "Tibet Leading China in Solar Energy Use," *Xinhua*, December 15, 2001, in FBIS-CPP20011215000038, which notes that this development "helps protect the fragile ecological system in the region."

[184] "APEC Workshop in Beijing," *Xinhua*, in *Alexander's* 5, no. 20 (November 1, 2000).

[185] *High-parametric* refers to a technology for increasing the efficiency of converting fuel to heat; *desulfurization* of coal allows it to be burned more efficiently and cleanly; *extra-high voltage transmission technologies* improve the efficiency of the power distribution system. "Energy at a Glance," accessed at <www.eia.doe.gov/emeu/cabs/china/eglance.html>.

[186] "China Set to Remain Asia's Hotspot for Oil and Gas Exploration," Reuters, in *Alexander's* 6, no. 20 (October 24, 2001).

[187] The World Bank classifies China as a "developing country" for the purposes of energy estimates. Foreign investment has played a critical role in financing the expansion of China's electric power infrastructure and is expected to play an even more important role in the future. Cited in <www.eia.doe.gov/oiaf/archive/ieo00/boxtext.html>.

[188] Phar Kim Beng gives the 1-month figure in "Oil Needs Drive China West," *Asia Times,* November 20, 2002, 2, accessed at <www.atimes.com/atimes/printN.html>. Liu Keyu of the Petroleum Economics and Information Center of CNPC projects stockpile goals of 20 days supply in 2005, 50 days in 2010, and 90 days in 2020 (VIC, "Asia Pacific Daily Summary," September 16, 2002).

[189] Peng Kailei, "China to Increase Strategic Oil Reserve," *Wen Wei Po* (Hong Kong), September 19, 2002, in FBIS-CPP2002091900000043; "China Plans Oil Stockpile," *China Daily*, accessed at <www.china.org.cn/english/2002/Sep/43358.htm>, cites a "government official from the State Council regarding concern about tension in the Middle East."

[190] "China Sets Principal Tasks for Upgrading of Petroleum Industry," *Xinhua*, September 27, 2002, in *Alexander's* 7, no. 20 (October 15, 2002), also lists priority areas for applying new technologies and priorities for research and development.

[191] "China's 'West Electricity for East' Program Begins in Sichuan," *Xinhua*, June 1, 2002, in FBIS-CPP20020601000004.

[192] "Japan, China Strike 5-Year Oil, Coal Trade Deal."

[193] "China's Energy Demand Now Exceeds Domestic Supply," DOE–EIA, accessed at <www.eia.doe.gov/emeu/cabs/china/part2.html#ENERGY>.

[194] Ibid.

[195] "China Sets Up Oil Company for Inland River Shipping," *Xinhua*, November 12, 2001, in FBIS-CPP20011112000176; "China to Construct Larger Oil Wharves to Meet Importation Demand," *Xinhua*, January 24, 2002, in FBIS-CPP20020124000045. Also see "Oil Wharf Upgrading Begins at Zhanjiang Port in South China," *Xinhua*, October 18, 2000, in FBIS-CPP20001018000203; "Two Companies Present LNG Carrier Designs to Chinese Shipyards," *Maritimepress*, in *Alexander's* 6, no. 20 (October 24, 2001).

[196] Liu Jipeng, "a power-reform expert" involved in drafting the national plan, quoted in Zhang Dong, "Power-Industry Reform Looms," *China Daily* (Hong Kong), September 10, 2002, in FBIS-CPP20020910000067.

[197] "China's Energy Conservation," *Xinhua*, in *Alexander's* 6, no. 24 (December 19, 2001), expresses Beijing's concern; Fu Jing, "State Stresses Importance of Energy Education," *China Daily*, December 17, 2001, in FBIS-CPP20011217000018, notes the government goal of "an annual 4 to 5 per cent energy-saving rate" and announces "massive publicity campaigns, education and training in this endeavor."

[198] "China Imports of Crude Have Almost Doubled," *Alexander's* 5, no. 21 (November 16, 2000); "China Reports on Crude Oil Imports," *Xinhua*, January 11, 2001, in *Alexander's* 6, no. 1, reports that China crude oil imports for the first 11 months of 2000 were 97 percent more than during the same period in 1999.

[199] "Marked Increase in Oil Exports to China," Iranian News Agency (Tehran), December 11, 2001, in FBIS-IAP20011211000038, also reported that Iran's exports in 2001 represented a 60 percent increase over the same period in 1999. "China and Ukraine Trying to Set Their Foot Strongly in Oman," *Times* (Oman), April 24, 2002, in *Alexander's* 7, no. 10 (May 16, 2002), reported that Sinopec intended expanding its activities in the country. This data in metric tons (2,204 pounds).

[200] "Beijing Posts Sharp Increase in Auto Sales," *Xinhua*, January 30, 2003, in FBIS-CPP20030130000071; "Fueling China's Growth," *The New York Times*, December 26, 2000.

[201] Manning, 104.

[202] Ibid., 105.

[203] "NPC Deputy Expects 150 Million Chinese Families to Buy Cars in Next 15 Years," *Xinhua*, March 12, 2003, quotes Chen Hong, "vice-president of Shanghai Automotive Industry Corporation Group." For environmental concerns, see "Beijing Encourages Use of Green Fuel," *Xinhua*, in *Alexander's* 7, no. 6 (March 21, 2002), which also reported that "the number of automobiles in Beijing has grown by more than 10 percent every year"; "China to Invest $120 mm in Developing Clean Motors," *Xinhua*, November 13, 2002.

[204] For instance, see "State Council Calls for Protection against Capricious Oil Market," January 2, 2001, accessed at <www.chinaonline.com/topstories/010102/1/B100122933.Asia>. Also see "PRC Researcher Notes Effect of Rocketing Oil Prices," China Internet Information Center, November 1, 2000, in FBIS-CPP20001102000026, which reports that the CNPC, Sinopec, and CNOOC all reported billions of dollars in losses for 1998 because of decreasing petroleum prices.

[205] "China's Largest Offshore Oilfield Found in Bohai Sea," *Xinhua*, February 1, 2000, in FBIS-FTS20000201000037.

[206] "China to Invest in Offshore Oil Exploration in Next Five Years," *Newspage*, and "China's Largest Offshore Oilfield to Start Producing in 2002," accessed at <www.Chinadaily.com.cn>, both in *Alexander's* 6, no. 1 (January 11, 2001): this report notes Phillips' share as $14.5 billion, an obvious exaggeration.

[207] "China to Close More Small Oil Refineries," AFP, cited in *Alexander's* 5, no. 7 (April 27, 2000), reports Beijing's concern about the number of small refineries that have resulted in an approximately 30 percent surplus in refining capacity.

[208] Chen Geng, director of the State Administration of Petroleum and Chemical Industries, quoted in "China to Take Measures to Control Petroleum Consumption," *Xinhua*, January 11, 2001, in *Alexander's* 6, no. 1 (January 11, 2001).

[209] See, for instance, Mai Tian, "Oil Reserves Plan in Pipeline," *China Daily*, January 20, 2003, in FBIS-CPP20030121000021; and "Chinese Government Builds Up Oil Stockpile," *Weweipo News*, February 25, 2003, in VIC, "Asia-Pacific Daily News Summary."

[210] "Chinese Government May Establish a 10-Billion Dollar Oil Fund," *China Daily*, March 3, 2003; "China to Build Two 2-Million Cubic Meter Strategic Oil Reserves," *China Times*, March 9, 2003.

[211] Niu Li, quoted in Xu Dashan, "Oil Reserve System Urgently Needed," *China Daily*, February 18, 2002, in FBIS-CPP20020218000024. See "China to Close More Small Oil Refineries" for a discussion of Beijing's decision to establish a strategic oil reserve. Sinopec president Wu Ruilin, quoted in Associated Press report, March 9, 2000, in *Alexander's* 5, no. 7 (April 27, 2000).

[212] Discussed in "Q & A on Oil Issues," *China Daily* (Hong Kong), November 20, 2000, in FBIS-CPP20001120000031. "Market Survey on Natural Gas Import From Siberia," *Xinhua*, November 25, 2002, in VIC, "Asia-Pacific Daily News Summary," notes that a "market survey will start in the near future" to assess the profitability of importing Russian natural gas.

[213] Chen Shihai, quoted in "China Plans to Launch State Strategic Oil Stockpile," *People's Daily Online*, in *Alexander's* 6, no. 15 (August 14, 2001). The 30-day estimate and cost figures are in Miccarelli, "China's Energy Future," 9.

[214] Song Wucheng, Ding Guosheng, and David Johnson, quoted in Gong Zhengzheng, "Analysts Urge State Oil Stockpile," *China Daily*, October 25, 2001, in FBIS-CPP20011025000042. Also see "China to Set Up Mineral Resources Reserve, Supply System," *Xinhua*, December 23, 2000, in FBIS-CPP20001223000056.

[215] "China Puts Off Fuel-Oil Futures Exchange Till 2003," *China Daily*, accessed at <www1.china daily.com.cn/news/cb/2002-10-28/91373.html>, reported that the China Securities Regulatory Commission is evaluating a proposal for this activity before making a recommendation to the State Council.

[216] Xie Ye, "Oil Future Market Mooted," *China Daily*, February 22, 2002, in FBIS-CPP20020222000030.

[217] "China Government Poised to Open Up Gas Market," *China Daily*, December 10, 2002, noted that this was accompanied by the qualification that local governments along the pipeline had to agree.
[218] "China to Abolish System of Fuel Oil Import Quotas by 2004," Reuters, cited in *Alexander's* 6, no. 1 (January 11, 2001), reports that less fuel oil was imported in 2000 than in 1999, although crude oil imports increased—which probably reflects greater or more efficient refining capacity in China. See Liu Ming, "Oil Industry Needs Reform for WTO Entry," *China Daily*, December 5, 2000, in FBIS-CPP20001205000013, for a discussion of how Beijing is addressing the question of import quotas because of pending entry into the World Trade Organization.
[219] "China Likely to Open Market for Refined Oil Products," *RiskCentre*, June 5, 2002, in *Alexander's* 7, no. 13 (June 27, 2002), discusses Sinopec joint ventures with Shell, ExxonMobil, and BP to operate gas stations in Jiangsu, Zhejiang, and Guangdong Provinces. Also see Xie Ye, "Sinopec Stock Issue Expected to Accelerate Breakup of State Oil Monopoly," *China Daily* (Business Weekly Supplement), November 5, 2000, in FBIS-CPP20001105000005; "China's Two Flagship Oil Companies Cast Oil Safety Line to Meet WTO Pledges," *Xinhua*, March 29, 2002, in FBIS-CPP20020329000114. Also see "Chinese Oil Majors to Reform Energy Sector Through Overseas Listings," Reuters, quoted in *Alexander's* 5, no. 20 (November 1, 2000).
[220] "China Shuts Down Illegal Gas Stations," *People's Daily*, in *Alexander's* 8, no. 2 (January 24, 2003), reports that almost 5,000 gas stations were shut down in 2002. "PetroChina Calls for Law to Regulate Gas-Station Franchising," *People's Daily*, July 31, 2002, in *Alexander's* 7, no. 16 (August 23, 2002), states that this company owns "about 2,000 franchised stations," while rival Sinopec owns 4,000.
[221] "Sinopec and PetroChina Still Dominate China's Gas Station Market," *SinoCast*, in *Alexander's* 8, no. 3 (February 6, 2003); "BP to Build 1,000 Petrol Stations in China," *Neftegaz.RU*, in *Alexander's* 8, no. 4 (February 20, 2003).
[222] "Experts Discuss Prospects of China's Oil Industry," *Xinhua*, November 1, 2001, in FBIS-CPP20011101000215. A different point of view is offered in Xie Ye, "China's Oil Imports Not to Shock Domestic Market After Ban Lifted," *China Daily*, November 13, 2001, in FBIS-CPP20011114000017.
[223] Xie Ye, "China to Open Oil Product Markets to Overseas Companies," *China Daily*, December 15, 2001, in FBIS-CPP20001215000022; and "China's Tightly-Controlled Refined Oil Market to Open Wider," *China Daily*, June 3, 2002, in FBIS-CPP20020603000021.
[224] "Import Quotas for Oil to Rise 15%," *China Daily*, August 2, 2002, accessed at <www.china daily.com.cn>.
[225] Peter Marber, "China's Energy Sector: Cheap But Don't Rush In," quoted in *Alexander's* 5, no. 21 (November 16, 2000), writing about the effects on China's energy sector of joining the World Trade Organization. Phillips, Exxon, Mobil, BP Amoco, and Shell were all early bidders.
[226] Chen Huai, "Active Role Needed in Oil," *China Daily*, July 7, 2002, in FBIS-CPP20020707000018. Chen, deputy director of the Market Economy Institute of the Development and Research Center of the State Council, urges China to be active in the futures market for petroleum products and establish "a huge reserve of oil forward contracts" with foreign oilfields.
[227] "China is Actively Exploring International Oil Markets," *Xinhua*, in *Alexander's* 4, no. 7 (July 19, 1999). Also see "Sudan: Chinese Electricity Firm Negotiating Projects with Energy Ministry," Suna News Agency, May 15, 2001, in FBIS-AFP20010516000021.
[228] "China's National Defense In 2002 'White Paper,'" December 9, 2002, 34, accessed at <www.china-embassy.org/eng/38991.html>.
[229] Ibid., 6.
[230] Yang Ron, "Perspective on Hot Spots," *Zhongguo Kongjun*, February 1, 1998, 4–6, in FBIS-FTS20000113001050, clearly envisions the United States as the would-be attacker.
[231] Philip T. Reeker, U.S. Department of State press statement, September 11, 2002. Also see Richard L. Armitage, "Statement at Conclusion of China Visit," Beijing, August 26, 2002.
[232] "Earnestly Strengthen Protection of Oil, Natural Gas Pipelines, Facilities," *Xinhua*, August 9, 2001, in FBIS-CPP20010809000116, notes that the original regulation was promulgated in 1989; this edition of *Xinhua* describes the revised regulation as "State Council Decree No. 313 of the PRC."
[233] "Four Sentenced to Death in Beijing for Destroying Power Lines," AFP, April 19, 2002, in FBIS-CPP20020419000058.
[234] "China is No Threat to America—For Now," *Guardian Unlimited*, in *Alexander's* 7, no. 9 (May 3, 2002), based partially on World Bank estimates.

[235] John Pomfret, "Chinese Oil Country Simmers as Workers Protest Cost-Cutting: Thousands Laid Off, Benefits Reduced," *The Washington Post*, March 17, 2002, A24, and Mark O'Neill, "Nearly 80,000 Daqing City Oil Workers Laid Off Since 2000," *South China Morning Post*, April 10, 2002, in FBIS-CPP20020410000048. Also, Robert J. Saiget, "PRC Factory, Oil Workers Protest in Northeast," AFP, March 18, 2002, in FBIS-CPP20020318000109, gives the figure of 80,000 laid off. According to Beijing, Daqing produced 51.5 million tons of crude oil in 2001, the 26th consecutive year of producing 50 million or more tons ("China's Oil Giant Reports High Output for 26 Straight Years," *Xinhua*, December 31, 2001, in FBIS-CPP20011231000120). For just two reports on the demonstrations, see "Further on 50,000 Oil Workers Protest," AFP, March 13, 2002, in FBIS-CPP20020313000158.

[236] "Curfew Imposed Following Demonstration Staged by 50,000 Workers," *Sing Tao Jih Pao* (Hong Kong), in FBIS-CPP20020320000066.

[237] See Larry Wortzel, "Beijing Struggles to 'Ride the Tiger of Liberalisation,'" *Jane's Intelligence Review*, January 1, 2001, for the best compilation of recent incidents of civilian unrest.

[238] This was reported in "Laid-Off Workers Demonstrate in Shanxi," *Information Center for Human Rights & Democracy* (Hong Kong), June 6, 2002, in FBIS-CPP20020607000038, 1. This report also stated that "30,000 workers in 20 factories in Liaoyang City, Liaoyang Province, staged a joint demonstration between 17 and 20 March."

[239] Statement by *Far Eastern Economic Review* correspondent, December 18, 2002. Reported in James Kynge, "Riots in Chinese Mining Town," *The Financial Times*, April 3, 2000, and "20,000 Go On Rampage Over Mine Closure," Associated Press, in the *South China Morning Post*, April 3, 2000; "APRC Riot Police Break Up Protest in Guiyang," AFP, March 29, 2000, in FBIS-CHI-2000-0329. These reports were brought to my attention by Dennis Blasko.

[240] "Jiang Zemin, A Report to CPC National Congress," November 8, 2002, accessed at <www.china.embassy-org/eng/index.html>.

[241] One associated mission in which the United States is playing only a peripheral role is that of combating piracy, an age-old problem in the region, most particularly in Southeast Asian waters; the nations of that region, however, have established an antipiracy organization that has precluded the necessity of the United States taking a leading role in the effort. This center and its activities are discussed in "Japan-China-S.E. Asia," *LLP Press*, May 4, 2000, accessed at <www.maritimesecurity.com/archive/aprilmay2000_anti_piracy_efforts.htm>.

[242] Among these border agreements reported in China's National Defense in 2002 "White Paper" are the "Beibu Gulf [Gulf of Tonkin] Demarcation Agreement with Vietnam" (December 2000), the "Supplementary Agreement on the [Tajik-PRC] Boundary" with Tajikistan (May 2002), frontier cooperation agreements with Russia (1995) and Mongolia, and a "Frontier Defense Cooperation Agreement" with Kazakhstan (January 2002). See "China and ASEAN Sign Spratlys Deal," Associated Press report, in *Alexander's* 7, no. 23 (November 27, 2002), for account of China and the ASEAN nations (Indonesia, Brunei, Singapore, Malaysia, Thailand, Vietnam, Laos, Cambodia, Burma, and the Philippines) signing a "Declaration on the Conduct of Parties in the South China Sea" designed to reduce the chances of military conflict over disputed territorial claims.

[243] "China, Kyrgyzstan Sign Anti-Terrorism Agreement," *Xinhua*, December 11, 2002, in VIC, "Asia-Pacific Daily News Summary."

[244] Wan Yanxian, Nian Hongtu, and Liang Yongli, "For Peace and Tranquility: A Talk with MG Liu Dengyun," *Jiefangjun Bao*, October 12, 2002, in FBIS-CPP20021012000012.

[245] First announced in "Shanghai Cooperation Organization Approves Center for Anti-terror," *China Daily*, June 8, 2002, accessed at <www.china.org.cn/english/FR/34120.htm>, but apparently not approved until the fall of 2002: see "Agree to Establish Antiterrorism Headquarters and Secretariat," *Xinhua*, November 23, 2002, accessed at <http://news.xinhuanet.com/newscenter/2002-11/23/content_638851.htm>.

[246] For instance, see Wang Jinguo and Yang Shu, "U.S. Enters Central Asia and the Caspian Sea for Oil and Gas," *Renmin Ribao*, May 13, 2002, in FBIS-CPP20020514000040.

[247] See Philip Andrews-Speed et al., *The Strategic Implications of China's Energy Needs*, Adelphi Paper 346 (London: International Institute of Strategic Studies, 2002), 35 ff., for a brief discussion of pipeline construction.

[248] "China Outlines Strategic Pattern of Natural Gas Development," *People's Daily*, in *Alexander's* 7, no. 23 (November 27, 2002), 1, which also included a CNPC report that China has signed a "protocol" with Russia for purchasing 20 billion cubic meters of natural gas annually, beginning in 2008.

[249] "China's 'Longest' Oil Pipeline Begins Trial Run," *Xinhua*, November 10, 2002, in FBIS-CPP20021110000058. By comparison, the Alaska pipeline is 800 miles long and cost $8 billion to build in 1977; the proposed Angarsk-to-Daqing line is projected to cost $2 billion, with another $5 billion required to develop the associated Russian gas deposits (Varvara Aglamishyan, "China Gives Go-Ahead for Deliveries of Russian Gas," *Nezavisimaya Gazeta*, October 23, 2002, 3, in FBIS-CEP20021024000180).

[250] Zhang Zhiheng, CCP Secretary in Xinjiang Autonomous Region, quoted in "China's Gas Line Is on the Way," *The Age Company*, June 6, 2002, in *Alexander's* 7, no. 13 (June 27, 2002). These security concerns are addressed at length in Peter S. Goodman, "A Pipeline to the Future, Clogged by China's Past?" *The Washington Post*, August 20, 2002, E1.

[251] Eight additional *Kilos* are on order from Moscow, and at least two more *Songs* are under construction. The 16 *Mings* and dozens of the *Romeos* remain seaworthy, but the PLAN will not be able to train enough crews to operate all of them effectively.

[252] The PLAN execution of the traditional naval mission of presence is detailed in Kenneth W. Allen and Eric A. McVadon, *China's Foreign Military Relations* (Washington, DC: The Henry L. Stimson Center, 1999); circumnavigation is addressed in many reports contained in "PLAN World Cruise 2002: Special Report," VIC, "Asia-Pacific Daily News Summary" (September 23, 2002).

[253] There is a large literature on this subject; see Samuel Eliot Morison, *The History of U.S. Naval Operations in World War II*, vol. I: *The Battle of the Atlantic* (Boston: Little Brown, 1947), for an account of the German navy's dramatically successful submarine campaign against coastal traffic off the eastern U.S. and Caribbean coasts in 1942 and 1943.

[254] Shigeo Hiramatsu, "China's Naval Advance: Objectives and Capabilities," *Japan Review of International Affairs* 8, no. 2 (Spring 1944), 126.

[255] Admiral Shi Yunsheng (commander of the PLAN), quoted in "Jiang Made the Final Decision on Adopting Offshore Defense Strategy," *Tung Fang Jih Pao* (Hong Kong), August 24, 2001, in FBIS-CPP20010824000062.

[256] Cited in "International Maritime Bureau Reports 57 Percent Increase in Pirate Attacks," *The Sun* (Vancouver), February 3, 2001, in *Alexander's* 6, no. 4 (February 22, 2001), 1; the $25 billion figure is in "Asia Piracy," Reuters (Singapore), December 10, 2002, in VIC, "Asia-Pacific Daily News Summary" (December 10, 2002). The most complete single source of information about the piracy problem is "Primer: Piracy in Asia," VIC (April 8, 2002).

[257] Discussed in "Pakistan, China to Ink Accord for Gwadar Port Construction," *The News* (Islamabad), August 8, 2002, in FBIS-SAP20010808000048.

[258] J. Mohan Malik, "Sino-Indian Rivalry in Myanmar," *Contemporary Southeast Asia* 16, no. 2 (September 1994), 137–155, presents an extreme view of a "de facto military alliance" between China and Burma. A less alarmist view is presented by William Ashton, "Chinese Bases in Burma—Fact or Fiction?" *Jane's Intelligence Review* 7, no. 2 (February 1995), 84–87. The author views the relationship as extensive but focused more on economic than military priorities, based on conversations with U.S., Chinese, Indian, and Taiwan analysts.

[259] "Indian Navy Exercises Seen to Irk Beijing," *The Washington Times*, May 8, 2000, 1.

[260] Ship numbers are from A.D. Baker III, ed., *Combat Fleets of the World 2002–2003* (Annapolis: Naval Institute Press, 2002). Reports have occasionally surfaced about India deploying nuclear-powered submarines—a *Charlie*-class was on loan from Russia for several years—but it is much more likely that both the Indian and Pakistani navies will restrict themselves to new submarines powered by air independent propulsion systems.

[261] For instance, see "Wang Shijie Says China Vows to Remain Active in Middle East Peace Process," AFP (Hong Kong), November 21, 2002, in FBIS-CPP20021121000024.

[262] See John H. Noer with David Gregory, *Chokepoints: Maritime Economic Concerns in Southeast Asia* (Washington, DC: National Defense University Press, in cooperation with the Center for Naval Analyses, 1996), for cost estimates.

[263] For a dissenting view, see Gordon G.G. Chang, *The Coming Collapse of China* (New York: Random House, 2001).

[264] The data used as the basis for this survey should be periodically updated if China's economy continues to expand, as expected. The following section draws on the excellent EIA report, "China's Energy Production and Consumption," in "China: An Energy Sector Overview, 1997" (December 1997), accessed at <www.eia.doe.gov/emeu/cabs/china/part2.html#ENERGY>; and EIA, "International Energy Outlook 2002, accessed at <www.eia.doe.gov/oiaf/ieo/tbl_bl.html>.

[265] "China's Modernization May be Slowed Down by Oil Shortage," *People's Daily*, in *Alexander's* 6, no. 15 (August 14, 2001).

[266] This estimate is from "World Energy Outlook 2000," International Energy Agency, accessed at <www.cna.ca/english/Articles/Electricity%20in%20China.pdf>, 2. Actual numbers offered by various Department of Energy, Chinese government, and commercial sources differ, but all agree on the magnitude of the expected increases in electrical production and consumption.

[267] "China Petrochemical Speeds Up Exploration in Northwest China," Reuters, in *Alexander's* 7, no. 3 (February 6, 2002).

[268] Tian Fengshan, cited in "China Spells Out New Strategy for Oil Exploration," *Xinhua*, April 18, 2002, in FBIS-CPP20020418000199, and "Land Minister on China's Potential in Oil, Gas Exploration," Xinhua, April 18, 2002, in FBIS-CPP20020418000157, also claimed that only 25 percent of China's onshore oil deposits and 20 percent of offshore deposits had been explored. A different opinion is given by Yao Guanming of the Chinese Academy of Engineering, who stated that the country has plentiful domestic reserves, with enough crude oil to be Acontinuously exploited" to 2063, and natural gas to the 22^d century ("Will China's Oil and Gas Resources Be Depleted," *Keji Ribao*, July 11, 2002, in FBIS-CPP20020729000142).

[269] *People's Daily*, February 12, 2003, in *Alexander's* 8, no. 5 (March 6, 2003), reported the ministry figures and repeated that 2002 oil imports were 15 percent above those for 2001.

[270] Yu Fei, "State Oceanic Administration: China to Launch Marine Satellite in 2001," *Xinhua* (Hong Kong), October 26, 2000, in FBIS-CPP20001026000072.

[271] See, for instance, Shigeo Hiramatsu, "Speculation Regarding the Appearance of Chinese Ships in Japanese Waters—Petroleum Resources Surveys or Military Strategy?" *Ekonomisuto* (Tokyo), March 19, 2002, in FBIS-JPP20020312000076. For a brief report of Beijing's diplomatic and economic activities in the South Pacific, see "Special Report: An Analysis of Chinese Activity in Oceania," VIC (November 5, 2002); Chinese aid to the region (approximately $300 million) has quadrupled since 1998. In the past few years, China has begun military contacts with Papua New Guinea, established a satellite tracking station in Kiribati, broken New Zealand's trade monopoly with the Cook Islands, and conducted numerous high-level political visits and exchanges in the region. Beijing has also secured observer (non-voting) status in the region's major economic organization, the Pacific Islands Forum (PIF). Former Vanuatu leader Barak Sope was directly responsible for securing China's position in the PIF.

[272] A good explanation of this promising resource is at <www.woodshole.er.usgs.gov/project-pages/hydrates/>. Also see "Broad Perspective of Gas Hydrates in Offshore Goa," accessed at <http://www.dod.nic.in/pro/chapter-16.doc>.

[273] "Gas Hydrates to be New Energy in 21st Century," *People's Daily*, December 15, 1999. Jin is also quoted in "Search for New Energy Resources has Become Imperative," *China Daily*, April 29, 2003, in *Alexander's* 7, no. 10 (May 16, 2003).

[274] "Jiang Zemin, Qadhafi Ink Deal to Open Libyan Oil Sector to PRC Firms," AFP, April 14, 2002, in FBIS-CPP20020414000057; Cao Xiaoxi, quoted in Xie Ye and Huo Yongzhe, "Oil Giants Map Out Overseas Takeovers," *China Daily* (Business Weekly Supplement), February 5, 2002, in FBIS-CPP20020205000060. Also see "CNPC Raises 2002 Production Forecast," Reuters, in *Alexander's* 7, no. 3 (February 6, 2002); "Sinopec Group and Sinochem to Buy Middle East Oil Assets," Reuters, in *Alexander's* 7, no. 8 (April 18, 2002), also reported that Sinopec has requested permission from Kuwaiti authorities to join in a $7 billion "Northern Fields" project. Also see "Oman and China Sign Oil Cooperation Agreements," *Asia Pulse*, in *Alexander's* 7, no. 8 (April 18, 2002). Sinopec's plan to invest $525 million in Algeria is reported in "Chinese Company Wins Contract," Radio Algiers, October 1, 2002, in FBIS-GMP20021001000257.

[275] One recent agreement with Venezuela is reported in "Foreign Minister Reports on Four Agreements Signed with PRC," Globovision Television (Caracas), December 28, 2001, in FBIS-LAP20011228000013.

[276] "CNPC Wants to Bid for Russia Slavneft," Reuters, December 4, 2002.

[277] "Yukos to Boost Oil Supply to China by Rail," *Commersant*, July 26, 2002, in *Alexander's* 7, no. 16 (August 23, 2002), reported these amounts as 11 million bbl in 2001, increasing incrementally to 22.1 million bbl in 2005.

[278] Cited in Ann Myers Jaffe and Steven W. Lewis, "Beijing's Oil Diplomacy," *Survival* 4, no. 1 (Spring 2002), 125.

[279] Huang Yan, president of PetroChina, quoted in "PetroChina May Invest in Russian Pipeline," Bloomberg.com, August 30, 2002, in *Alexander's* 7, no. 18 (September 19, 2002).

[280] "Russia and China May Ink Oil Pipe Deal Next Month," Reuters, November 20, 2002; "Government to Approve a China-Russia Oil Pipeline Project Soon," *Wenweipo News*, in VIC, "Asia-Pacific Daily News Summary" (November 25, 2002).

[281] Huo Yongzhe, "Pipe Dream to Come True for Oil/Gas Transfer," *China Daily* (Business Weekly Supplement), September 10, 2002, in FBIS-CPP20020910000086.

[282] An optimistic account of these negotiations is in Alexandre Y. Mansourov, "Russian-Chinese Strategic Rapprochement: Lessons of History and Outlook for the New Millennium," *Washington Journal of Modern China* 7, no. 2 (Fall-Winter 2001/2002), 29. The $2 billion figure is in Alexandre Zyuzin, "Russia-China Oil Pipeline Construction to Boost Bilateral Trade," ITAR–TASS, December 1, 2002, in FBIS-CEP20021201000008, which reports that pipeline construction will start in 2003 and be completed in 2005, but also notes that the project's "feasibility study" remains uncompleted. "Assessing China-Russia Oil Pipeline Project," *Takungpao News*, December 5, 2002 in VIC, "Asia-Pacific Daily News Summary," reports that the pipeline feasibility study had just begun.

[283] A good summary of this imbroglio is in "Russia's Potential Far East Pipelines," VIC, "Special Report: Potential Russian Far East Oil Pipelines" (February 20, 2003), accessed at <www.oilandgasinternational.com/departments/development_production/jan03_russia>, but dozens of reports in Russian, Chinese, and Japanese newspapers have appeared on these pipeline negotiations. See, for instance, "Russian Minister Views Political Aspect of Oil Pipeline Projects," ITAR–TASS, February 28, 2003, in FBIS-CEP20030228000017; Wang Ling, "China May Approve Adding Japan to Sino-Russian Oil Pipeline," *China Daily*, March 3, 2003, in FBIS-CPP20030303000025; and "Russian President Favors Japan Pipeline Plan Over China's," *Kyodo News*, January 30, 2003, accessed at <www.asia.news.hahoo.com/030130/kyodo/d7oseeqo2.html>. Also see "Russia-China Oil Pipeline Project Stalled Because of Internal Politics," *Neftegaz.RU*, December 24, 2002, in *Alexander's* 8, no. 2 (January 24, 2003), for the probably accurate report that "the major obstacles for the crucial project . . . have become Russia's high prices and China's demands for a share in Siberian oilfields." The dual-line solution is described in "Russian Oil Pipeline Confirmed to Lead Directly to China," *People's Daily*, March 18, 2003, despite this article's misleading title.

[284] See, for instance, "Creation of ASEAN Gas Centre Expected to Take Off in 2002," *Manila Bulletin*, in *Alexander's* 6, no. 24 (December 19, 2001); Manning, 113.

[285] "Southeast Asia Needs $180 Billion Investments to Develop its Energy Sector," *The Daily Star*, in *Alexander's* 6, no. 5 (March 8, 2001).

[286] "Japan, China, and South Korea Hold High-Level Energy Talks," *Kyodo News*, in *Alexander's* 7, no. 6 (March 21, 2002).

[287] "Asia Will Need $200 billion in Next 10–15 Years to Build Gas Infrastructure," Gulf News Online, May 22, 2002, in *Alexander's* 7, no. 12 (June 13, 2002).

[288] "Asian Countries Agree on Coordinated Response to Oil Supply Emergency," *Xinhua*, September 22, 2002, in *Alexander's* 7, no. 20 (October 15, 2002). The meeting was attended by 65 nations, plus 10 international organizations; 13 states agreed to form the information-sharing network. This meeting is also described in "Japan, 12 Other Asian Nations Agree on Oil Supply Info Framework," *Jiji Press* (Tokyo), September 22, 2002, in FBIS-JPP20020922000059.

[289] General Office of the State Council, "The Suggestions on Further Rectifying and Standardizing the Market Order of Finished Oil," *Xinhua* (Hong Kong), October 25, 2001, in FBIS-CPP20011025000151. Also see "Refined Oil Prices Drop Again Today," *Xinhua*, December 31, 2001, in FBIS-CPP20011231000160.

[290] Xia Qiuju, "State-Owned Oil Company in Liaoning Makes Progress in Restructuring SOEs," *China Daily*, June 10, 2002, in FBIS-CPP20020610000014.

www.ingramcontent.com/pod-product-compliance
Lightning Source LLC
Chambersburg PA
CBHW051336170526
45166CB00002B/844